CAMBRIDGE COUNTY GEOGRAPHIES

General Editor: F. H. H. GUILLEMARD, M.A., M.D.

RUTLAND

Cambridge County Geographies

RUTLAND

by

G. PHILLIPS

Editor of the *Rutland Magazine and County Historical Record*

With Maps, Diagrams and Illustrations

Cambridge:

at the University Press

1912

CAMBRIDGE UNIVERSITY PRESS
Cambridge, New York, Melbourne, Madrid, Cape Town,
Singapore, São Paulo, Delhi, Mexico City

Cambridge University Press
The Edinburgh Building, Cambridge CB2 8RU, UK

Published in the United States of America by Cambridge University Press, New York

www.cambridge.org
Information on this title: www.cambridge.org/9781107696419

First published 1912
First paperback edition 2013

A catalogue record for this publication is available from the British Library

ISBN 978-1-107-69641-9 Paperback

PREFACE

IN addition to the acknowledgements made in the text I wish to thank Miss Tryon of Hambleton for botanical notes, and Miss Wingfield of Market Overton and Mr W. N. Wortley of Ridlington for information on rainfall.

I have also to acknowledge the great assistance obtained from contributions to the *Rutland Magazine* by various authors, and especially from the late Mr R. P. Brereton's article on the characteristics of Rutland Churches.

G. PHILLIPS.

Nov. 1912.

CONTENTS

ILLUSTRATIONS

The illustrations reproduced on pp. 12, 15, 29, 48, 94, 101, 113, 121, 135, 155, 156, 162, 164, and 165 are from photographs by Mr W. J. W. Stocks, of Uppingham; those on pp. 74 and 134 are from old prints. That on p. 81 was kindly supplied by Mr V. B. Crowther-Beynon, F.S.A., and the author is indebted to the Secretary of the Society of Antiquaries for permission to reproduce the illustration on p. 87. All other illustrations are reproduced from photographs taken by the author.

1. County and Shire. The word *Rutland*: its Origin and Meaning.

Our land, as we all know, is divided up into counties and shires. Nowadays these terms mean much the same thing, but this was not so at one time, and indeed is not so now, accurately speaking. In order to understand the terms we must first turn to their origin and meaning.

The shire as a division of land has been in existence from very early times. It is often stated that King Alfred (A.D. 871–901) made England into shires, but this is hardly correct, for some are spoken of as such in the early chronicles long before his date, while the county of London, on the other hand, which is essentially a shire, only came into being in 1888. The shire is really the *share* or part *shorn* off (for the words have a similar derivation from the Anglo-Saxon *scir*, *sciran*, to divide, to cut) from some previously-existing state. Thus the great mid-England kingdom of Mercia was parted into many shires in ancient days. Other kingdoms, however, from geographical or other reasons, have altered but little, and have retained the same boundaries and almost the same names for more than a thousand years, as we see in the case of Essex, Kent, and Sussex—the kingdoms of the East Saxons, the Cantii, and the South Saxons respectively.

Let us turn to the word county. This is of much later origin, for it dates only from the time when William the Conqueror parted the land for administrative purposes among the great nobles who came over with him. The county (Lat. *comitatus*) was the area assigned to the Count (Lat. *comes*, French *comte*), the companion, in other words, of the King, and though the title of Count has failed to establish itself in the place of that of Earl, we still retain "Countess" and "county" as English words.

The early chroniclers—the Saxon Chronicle, Simeon of Durham, Florence of Worcester, and William of Malmesbury—while referring to its surrounding neighbours, make no mention of Rutland; and, although the name appears in Domesday Book, it has quite a modern history as a shire compared with Leicestershire, Nottinghamshire, and Northamptonshire, and appears to have been carved out of one or more of these counties.

But how and when Rutland became a shire is a question of which the solution is by no means easy, although much attention has been given to it by students of Domesday Survey. This much, however, is certain, that the name of Roteland (Rutland) is earlier than the Norman Conquest, and the portion of land known by that name was not a shire at the Domesday Survey. Probably the name, as well as the district, is far older than the division of Mercia into shires by Edward the

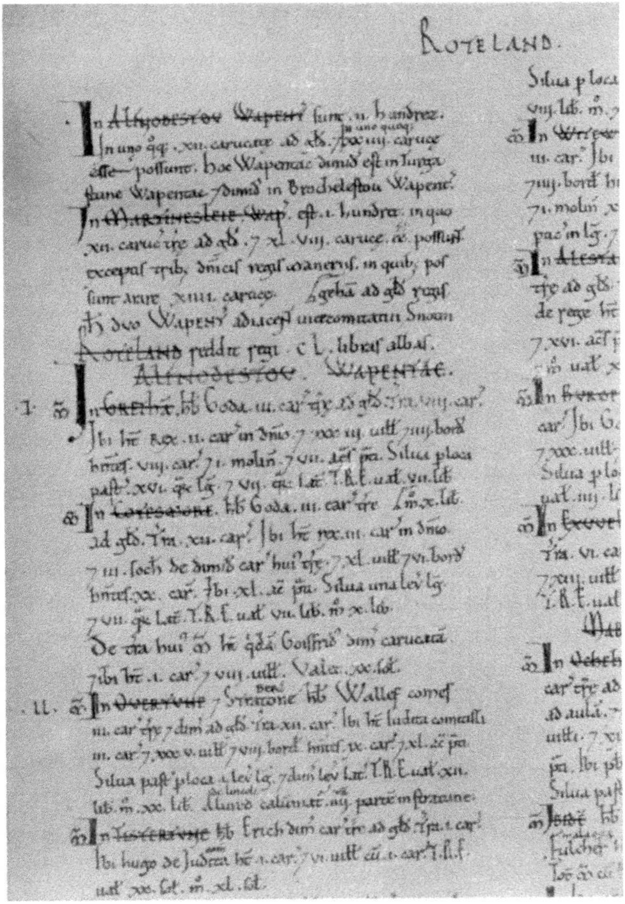

Portion of Domesday Book relating to Rutland

Elder, for it belongs to a class common to the North and Midlands like Holland (the fen-land area of Lincolnshire), Cleveland (the North Riding district), Westmorland, and Cumberland, representing the old native division of the soil prior to the Danish Conquest. Not one of these was a shire at the time of Domesday ; but Westmorland, Cumberland, and Rutland became so later, while Cleveland and Holland remain mere popular names to the present day.

No one in Rutland, of course, would ever speak of " Rutlandshire," any more than people would speak of " Cumberlandshire " and Westmorlandshire " in those counties. Everything indeed seems to show that the district, as a popular division, goes back to a far earlier time than the artificial arrangement which made it into a recognised administrative unit. One mark of its real origin may, perhaps, be seen in the fact that, alone among Mercian shires, it is not named after its county town, otherwise it would have been called Oakhamshire. Apparently, as a recent writer has said, it remains a solitary example of an old Mercian division which has outlived the West Saxon redistribution of the country into shires, rudely mapped out around the chief Danish burghs. In this connection it is interesting to note that Danish local names are unknown in the county, and that the subdivisions of the soil, though sometimes described by their Scandinavian appellation of wapentakes, are far oftener designated in the true old English style of hundreds.

According to one early authority Rutland was given

by Ethelred to Emma, his Norman Queen, on his
marriage in 1002. This was the beginning of that
connexion of Rutland with the queens and the favourites
of the kings of England which forms the main interest
of its story. Most probably, on the reorganisation of the
Mercian shires after their reconquest from the Danes,
the whole of the district formed part of Northamptonshire, and it was the giving of Rutland to successive
queens, as a dower, that caused it ultimately to be formed
into a separate county.

Edward the Confessor followed the example of his
father by granting Rutland to his queen, Edith, whose
name still lingers to the present day in the village of
Edith Weston. He afterwards granted it to Westminster
Abbey, reserving a life interest to the Queen, but
William the Conqueror refused to confirm the grant,
and on the death of Edith in 1075, took it into his own
hands. For several centuries the barony of Oakham
with the county or shrievalty of Rutland remained to
the Crown as a valuable possession with which to endow
members of the Royal house or to secure or reward the
services of its supporters.

The origin of the name Rutland has puzzled many
topographical writers. We may dismiss at once the
fable that a certain Mercian king—whose name, by the
way, is not mentioned—having a favourite named Rut
or Roet, gave him "as much land in this part of his
kingdom as he could ride round in a day, and he, riding
about the land now made into a county within the time
appointed, had it therefore given him, and he imposed

upon it the name of Rut's land, now for brevity's sake called Rutland." The above, however, is not more astonishing than the theory that Rutland was so called from its circular shape, *quasi Rotundalandia*, as if our ancestors usually spoke bad medieval Latin ; or from

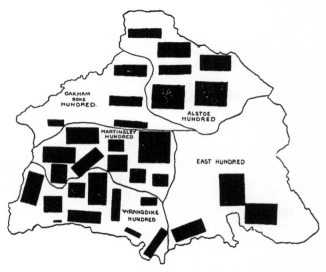

Map showing proportion of red land in Rutland

roet, the old Romance word for a wheel, as if they spoke Norman French in the days of Alfred and Athelstan.

The old county historians who gave us Rut and Rotundalandia as the origin of the name discarded Rud or Redland (from the ruddy complexion of the soil), because, said one, from observation there was only one small part of the county where red land predominated. This point has, however, been settled by the present writer, for it is a fact that no less than 24,178 acres out of 97,073 acres, or about one-fourth of the soil of the county, is known as red land, and it is scattered about in as many as 32 out of the 57 parishes which the county contains.

In the map here shown the dark portions denote the number of acres of red land in the respective parishes. A glance will show how this is distributed, and it will be noted that the larger portion, estimated at 15,332 acres, in the Martinsley, Oakham Soke, and Alstoe hundreds, is situated in that part of the county which is recognised by all authorities as the district termed Roteland, before the additions were made which come within the present shire boundary. There is ample evidence therefore, that this colour would be a predominating feature of the landscape.

2. General Characteristics.

Situated nearly in the heart of England, among the lowlands which slope downward to the fen country, lies the little agricultural county of Rutland—a pigmy among the giants—for it is encircled by shires whose areas run to five, six, and even eighteen times its acreage.

In many a part of the three kingdoms there may be found grander views and bolder scenery, but in none can there be seen more beautiful landscapes of softly blended wood, water, and pastoral country than are to be found in our county. The frequent alternations of hard and soft beds of rock has given rise to a considerable variety of scenery. This, and the gentle and regular dip of the beds have been productive of the numerous flat-topped hills which are very characteristic of this part of England.

There is no river of any importance in Rutland, but the Welland, Gwash or Wash, and the Chater, each running through richly-cultivated and densely-wooded districts, afford some delightful miniature river scenery.

The wide and extensive views to be obtained from Manton, Brooke, Barnsdale, Market Overton, Wardley, Preston, Whissendine, Ranksborough, Stoke Dry, and other points of vantage, make it an ideal country for the lover of fox-hunting. Hills clothed from base to summit with majestic trees and luxuriant evergreens, sunny dells and silvery streams, fertile vales and swelling uplands, add a charm to Rutland, notwithstanding its small size, which did not escape the eye of the poet Michael Drayton, who was probably staying at Ridlington when he wrote his description of this part of the county. In the *Polyolbion*, a poetical description of England, published in 1613–22, the following lines appear:—

"Love not thyself the less, although the least thou art,
  What thou in greatness want'st, wise nature doth impart
  In goodness of thy soil; and more delicious mould,
  Surveying all this isle, the sun did ne'er behold.
  Bring forth that British vale, and be it ne'er so rare,
  But *Catmus* with that vale for richness may compare.

What forest nymph is found, how brave soe'er she be;
But *Lyfield* shews herself as brave a nymph as she.
What river ever rose from bank or swelling hill,
Than Rutland's wandering Wash, a delicater rill?
Small shire that can'st produce to thy proportion good,
One vale of special name, one forest, and one flood!
Oh! *Catmus* thou fair vale, come on in grass and corn!
That *Beaver*[1] ne'er be said thy sisterhood to scorn,
And let thy *Ockham* boast to have no little grace,
That her the pleased Fates did in thy bosom place!
And *Lyfield*, as thou are a forest, live so free,
That every forest nymph may praise the sports in thee;
And down the *Welland's* course, Oh! *Wash*, run ever clear.
To honour, and to be much honoured by this Shire."

The south-western part of the county was formerly entirely occupied by the Forest of Leighfield—one of the royal forests of England—part of which still remains, including Beaumont Chase, and Burley, Exton, and Normanton Parks; the two latter, still dotted by herds of deer, occupy no inconsiderable fraction of its area.

A magnificent landscape spreads before us as we pass through the rich and beautiful Vale of Catmose, which, running from Ranksborough Hill on the western boundary to the centre of the county, includes within its limits the county town of Oakham.

On the north-eastern side of the Vale, the elevated ground, beginning at Burley-on-the-Hill, forms a level tableland and stretches all over the northern part of the county, overlooking the fertile and well-wooded plains of Leicestershire and Lincolnshire.

[1] Belvoir Vale, Notts.

On the breezy tableland which overlooks the Eye and the Welland, and forms part of the long sloping wolds which roll away through the neighbouring shires, stands the only other town in Rutland—Uppingham—famous for its School, the buildings of which cover no small portion of the town which Leland, writing in 1545,

The Vale of Catmose

rather contemptuously described as consisting of " but one meane strete and but a very meane church."

Rutland is essentially an agricultural county. No belching smoke-stacks poison the air, and we need not walk far before the silence is broken only by the hum of the insect world, which reminds us that we are far removed from busy loom and spindle. The county has not been invaded by the modern factory. The primitive picturesque thatched houses, to be found even in the county town, are characteristic of the district ; they add charm to the view, and mark a time when modern blue slates were not available. The villagers still retain many of the primitive manners of their forefathers, but the smock frock of the agricultural labourer has vanished.

The exigencies of rapid transit have, indeed, left a mark on Rutland, as on other agricultural counties, for the character of farming has changed within the last fifty years to such an extent that arable land has disappeared by ten thousand acres and given place to pasture.

As mentioned in the previous chapter, the predominating colour of the ploughed land is red. " Rutland raddlemen " has become a proverb, a nickname given, long ago, to the inhabitants of this county, very much in the same way as its neighbours are designated " Lincolnshire bagpipes."

3. Size. Shape. Boundaries.

Rutland, while classed as one of the Midland Counties of England, occupies a position more easterly than central, being separated from the Wash only by the narrowest part of Lincolnshire.

The county's breadth and length are nearly equal, measuring a little over 17 miles. Its greatest breadth is from the eastern side near Essendine, on the borders

of Lincolnshire, to the westernmost point, near Whissendine, on the borders of Leicestershire. Its greatest length
is from near Thistleton in the north, to where the Eye
Brook falls into the Welland in the south, near Caldecott.
The circumference is 63 miles, and the area 97,273 acres,
or approximately 152 square miles. In size it is the least
of all the English counties and of all the Welsh except
Flint.

The Welland Valley, from Rockingham

In shape it may be described as roughly triangular,
with a western, a northern, and a south-eastern side,
the contour of all the sides being irregular. It bears
a distinct resemblance to the map of Spain and Portugal.

Rutland is bounded on the east and north-east by
Lincolnshire, on the west and north-west by Leicestershire,
and on the south-east by Northamptonshire. By reference to the map it will be seen that the town of

Stamford, or at least so much as lies north of the river Welland, is included in Lincolnshire. The Welland being a natural boundary, the almost rectangular piece on which the Royal Borough stands might excusably be expected to belong to Rutland and how it escaped inclusion within its boundaries history does not say. Possibly, however, the explanation lies in the fact that under the Anglo-Saxon rule the counties of Lincoln, Northampton, and Rutland formed part of the kingdom of the Mercians, whose king, Wolfhere, completed the building of the church of Medehamstede (Peterborough Abbey), and by charter confirmed to it, in 664, "all that part of the town of Staunford which is towards Medehamstede beyond the bridge." Probably this, then, is the reason why this strip of land was included in Lincolnshire, for the parish of St Martin's, called Stamford Baron, is in the county of Northampton.

4. Surface and General Features.

The physical features of the county are entirely dependent upon its geological formation. This, as we shall see later, has also determined the situations of the towns and villages, as well as the occupations and the industries of the inhabitants.

The alternation of beds of varying hardness, as clay, ironstone, sandstone, and limestone, has produced, as already stated, much variety of scenery. The softer beds having been more rapidly worn away by varied agencies, such as rain, rivers, and frost, are now found as low plains, as may be seen in the valley of the Welland and the Vale of Catmose. The harder beds stand out prominently as hills, plateaus, and escarpments, or are seen as shelf-like terraces on the sides of the valleys. These latter features are especially prominent in the west of the county, where the great thickness of the Upper Lias clays, reaching in places some 200 feet, has given rise to scenery comparable in boldness to that of the Cotswold Hills.

The surface is finely varied with gentle elevations and depressions ; the elevations generally running east and west, divided by valleys about half-a-mile in width. Among these are the extensive open valley forming part of the basin of the Welland on the south-eastern side of the county, and the rich and beautiful Vale of Catmose running from the western side.

The soil is various, but generally fertile. That of the eastern and south-eastern districts is mostly of a shallow staple, on limestone rock ; but in nearly all other parts a strong red loam, resting on a substratum of blue clay, prevails. The red soil and several chalybeate springs indicate the existence of iron, which is being worked in the Cottesmore and Market Overton districts. The iron occurs in a bed called by geologists the Northampton Sand (see geological section) which occupies much of the surface.

The woods of Rutland are supposed to have been formerly very much more extensive than at the present. A writer, about 100 years ago, estimated them at 2815 acres, and some authors have asserted that the whole of the Vale of Catmose was once an extensive tract of woodland. The word Catmose, in fact, is thought by some to be probably derived from *Coet-Maes*, the Celtic for a woody plain.

The forest of Leighfield, or Lyfield, once occupied the greater part of the Oakham Hundred ; and that of

Castle Hill, Uppingham

Beaumont Chase, a part of it, extended over a large area of Martinsley Hundred and had several villages within its confines, though they are now destroyed. Several parishes in the vicinity still claim forest rights. Leland, when speaking of this forest in the reign of Henry VIII, says :—
" from Wiscombe partly through woddy ground and the forest of Leafield, and so on to Rutlandshire by woddy first, and then by champain ground, but exceedingly rich

of corn and pasture." This description may be considered as strictly applicable at the present day, as there is not another tract of the same extent in England which presents a richer prospect of wood and cultivation than this when viewed from the rising ground between Uppingham and Wardley.

5. Watershed. Rivers.

No rivers of any importance, if we except one of the head-streams of the Lincolnshire Witham, rise in Rutland; indeed, its rainfall being small, its soil porous, and its hills of only moderate height, the comparative amount of water which runs over its surface is probably less, area for area, than that of any other county in England.

The Welland with its tributaries the Chater and Gwash are the only streams of any size; and of these, the main stream, the Welland, forms the boundary of the county for a considerable distance on its south-eastern side. The course of these streams from west to east, across the edges of the strata, cutting through a succession of hard and soft beds, seems at first sight rather exceptional. We might have expected them to run in long narrow valleys from north to south, having scooped out ravines in the clay beds, while the harder strata would hem them in on each side. Such is, in fact the direction and situation of the later-formed streams, as may be well seen by walking up the Eye Brook by Stockerston and Allexton, or near the source of the other river Eye in the north. The main rivers, however, as the

Welland, had the direction of their course determined in times very far back, when the contour of the surface was quite different from what it is now. Higher beds of rock, the chalk now denuded off, were then continuous to the westward, and the rivers flowed on the surfaces of beds of tolerably equal hardness beneath, being kept in their courses by the surface strata.

The Welland at Barrowden

Since then, these upper rocks have been removed, and we now have the apparent anomaly of streams cutting great gashes in escarpments of limestone and ironstone, as the Welland has done between Harringworth and Morcott, the Chater between North Luffenham and South Luffenham, and the Gwash at Empingham. The same reasons apply elsewhere; as, for instance, in the extraordinary manner in which the Thames cuts through the chalk escarpment at the Goring Gap near Reading, instead of turning to the north or south, as one would expect it to do.

At least 130 out of the 150 square miles comprised in the county belong to the drainage basin of the Welland. This river rises near Husbands Bosworth in Leicestershire. For the greater part of its course it is a boundary river, separating Leicester, Rutland, and Lincoln successively from the county of Northampton. Beginning at the confluence of the Eye, by Rockingham Station near Caldecott, it passes Thorpe-by-Water, Seaton Mill, and Barrowden, then taking the south side of Tixover Church and past Tixover Hall it skirts Duddington (Northants.) and Tixover Grange, and passes under Collyweston Bridge to Tinwell; then, at Broadeny Bridge, the Rutland boundary ends. The river continues its course through Stamford, and finally, crossing the Fen district, discharges into the Wash.

The Eye Brook, or Little Eye river, rises on Tilton Hill in Leicestershire, and for some miles forms the boundary of the county, beginning at Finchley Bridge near Belton. It takes a south-easterly course, passing north of Allexton (Leicestershire) and south of Beaumont Chase to Stoke Dry, whence it follows the valley to Caldecott and joins the Welland at Rockingham Station.

The Chater has its source on Whadborough Hill, also in Leicestershire, and close to Tilton Hill. It enters the county at Leighfield, passes Ridlington on the north, follows the valley between Manton and Preston, continues its sinuous course between the two Luffenhams and through Ketton, and finally joins the Welland near Tinwell, two miles west of Stamford. It is 15 miles in length.

The Gwash also rises in Leicestershire, near Burrough Hill. It takes an easterly course. Passing through Braunston and south of the Oakham Waterworks to

The Gwash at Ryhall

Brooke, it runs under the Midland Railway line below Gunthorpe, skirts Hambleton wood, and continues through Brake Spinney to Normanton ponds. Here, just before entering Mow Mires Spinney, it receives, as one of its principal tributaries, the water from Burley Ponds. At Empingham Mill its volume of water is still further increased by that which flows through the ponds at Exton

Park ; and at Wild's Lodge by the springs in Shacklewell Spinney.   Continuing a most erratic course through Tickencote, the river crosses the Great North Road under the Roman Bridge at Great Casterton.   After touching Little Casterton and Tolethorpe it nearly encircles the village of Ryhall, where it runs down the side of the main street, and passing through Belmisthorpe, finally discharges into the Welland below Newstead Mill beyond Stamford, a course of about 25 miles.

The River Glen, a Lincolnshire tributary of the Welland, passes for about a couple of miles through the north-eastern corner of the county, and joins the latter stream in the Fens.

It only remains to mention another watercourse, known as the Whissendine brook. This is a feeder of the River Eye, running from Cold Overton Lake through Stapleford Park, and is celebrated in the annals of the chase for the many exciting scenes it has produced when hounds have been in full cry.

6. Geology and Soil.

By Geology we mean the study of the rocks, and we must at the outset explain that the term rock is used by the geologist without any reference to the hardness or compactness of the material to which the name is applied ; thus he speaks of loose sand as a rock equally with a hard substance like granite.

Rocks are of two kinds, (1) those laid down mostly under water, (2) those due to the action of heat.

The first kind may be compared to sheets of paper laid one over the other. These sheets are called beds, and such beds are usually formed of sand (often containing pebbles), mud or clay, and limestones, or mixtures of these materials. They are laid down as flat or nearly flat sheets, but may afterwards be tilted as the result of movement of the earth's crust, just as we may tilt sheets of paper, folding them into arches and troughs, by pressing them at either end. Again, we may find the tops of the folds so produced worn away as the result of the wasting action of rivers, glaciers, and sea-waves upon them, just as we may cut off the tops of the folds of the paper with a pair of shears. This has happened with the ancient beds forming parts of the earth's crust, and we therefore often find them tilted, with the upper parts removed.

The other kinds of rock are known as igneous rocks. With these we in Rutland are not concerned.

The production of beds is of great importance to geologists, for by means of these beds we can classify the rocks according to age. If we take two sheets of paper, and lay one on the top of the other, the upper one has been laid down after the other. Similarly with two beds, the upper is also the newer, and the newer will remain on the top after earth-movements, save in very exceptional cases which need not be regarded by us here, and for general purposes we may regard any bed or set of beds resting on any other in our own country as being the newer bed or set.

The movements which affect beds may occur at

different times. One set of beds may be laid down flat, then thrown into folds by movement, the tops of the beds worn off, and another set of beds laid down upon the worn surface of the older beds, the edges of which will abut against the oldest of the new set of flatly-deposited beds, which latter may in turn undergo disturbance and renewal of their upper portions.

Again, after the formation of the beds many changes may occur in them. They may become hardened, pebble-beds being changed into conglomerates, sands into sandstones, muds and clays into mudstones and shales, soft deposits of lime into limestone, and loose volcanic ashes into exceedingly hard rocks. They may also become cracked, and the cracks are often very regular, running in two directions at right angles one to the other. Such cracks are known as joints, and the joints are very important in affecting the physical geography of a district.

If we could flatten out all the beds of England, and arrange them one over the other and bore a shaft through them, we should see them on the sides of the shaft, the newest appearing at the top and the oldest at the bottom, as in the annexed table. Such a shaft would have a depth of between 50,000 and 100,000 ft. The strata beds are divided into three great groups called Primary or Palaeozoic, Secondary or Mesozoic, and Tertiary or Cainozoic, and the lowest of the Primary rocks are the oldest of Great Britain, and form as it were the foundation stones on which the other rocks rest. These are termed the Pre-Cambrian rocks. The three great

	Names of Systems	Subdivisions	Characters of Rocks
TERTIARY	**Recent** **Pleistocene**	Metal Age Deposits Neolithic ,, Palaeolithic ,, Glacial ,,	Superficial Deposits
	Pliocene	Cromer Series Weybourne Crag Chillesford and Norwich Crags Red and Walton Crags Coralline Crag	Sands chiefly
	Miocene	Absent from Britain	
	Eocene	Fluviomarine Beds of Hampshire Bagshot Beds London Clay Oldhaven Beds, Woolwich and Reading Thanet Sands [Groups	Clays and Sands chiefly
SECONDARY	**Cretaceous**	Chalk Upper Greensand and Gault Lower Greensand Weald Clay Hastings Sands	Chalk at top Sandstones, Mud and Clays below
	Jurassic	Purbeck Beds Portland Beds Kimmeridge Clay Corallian Beds Oxford Clay and Kellaways Rock Cornbrash Forest Marble Great Oolite with Stonesfield Slate Inferior Oolite Lias—Upper, Middle, and Lower	Shales, Sandstones and Oolitic Limestones
	Triassic	Rhaetic Keuper Marls Keuper Sandstone Upper Bunter Sandstone Bunter Pebble Beds Lower Bunter Sandstone	Red Sandstones and Marls, Gypsum and Salt
PRIMARY	**Permian**	Magnesian Limestone and Sandstone Marl Slate Lower Permian Sandstone	Red Sandstones and Magnesian Limestone
	Carboniferous	Coal Measures Millstone Grit Mountain Limestone Basal Carboniferous Rocks	Sandstones, Shales and Coals at top Sandstones in middle Limestone and Shales below
	Devonian	Upper } Devonian and Old Red Sand- Mid } stone Lower }	Red Sandstones, Shales, Slates and Lime- stones
	Silurian	Ludlow Beds Wenlock Beds Llandovery Beds	Sandstones, Shales and Thin Limestones
	Ordovician	Caradoc Beds Llandeilo Beds Arenig Beds	Shales, Slates, Sandstones and Thin Limestones
	Cambrian	Tremadoc Slates Lingula Flags Menevian Beds Harlech Grits and Llanberis Slates	Slates and Sandstones
	Pre-Cambrian	No definite classification yet made	Sandstones, Slates and Volcanic Rocks

groups are divided into minor divisions known as Systems. The names of these systems are arranged in order in the table, and the general characters of the rocks in each system are also stated.

With these preliminary remarks we may now proceed to a brief account of the geology of Rutland.

As we go from west to east in England we pass generally from older to newer rocks.

In Rutland, the rocks which come to the surface and form the subsoils belong to two different periods, separated by an immense interval of time. These are the rocks of the Jurassic system and those of the late Tertiary or Post-Tertiary period.

Before proceeding further it would be well to clearly indicate the order of succession of the rocks in Rutland.

RECENT AND PLEISTOCENE.

Post-Glacial Period.	River gravels and alluvial deposits.
Glacial Period.	Boulder clay and gravels.
Pre-Glacial Period.	Sands and clays.

JURASSIC.

Cornbrash.

Great Oolite.
- Great Oolite clays.
- Great Oolite limestone.
- Upper estuarine clays.

Inferior Oolite.
- Lincolnshire Oolite limestones.
- Northampton sand.

Lias.
- Upper Lias clay and limestone.
- Marlstone rock.
- Middle Lias clay.
- Lower Lias clay.

The total thickness of the Jurassic series exposed in Rutland is about 1200 ft., but this does not hold good over any considerable area, as the beds thin out in some directions and thicken in others.

Quite distinct from the Jurassic rocks, and, as we have said, marked by a long interval of time, are the Post-Tertiary or Glacial deposits of clay, gravel, and sand, which are spread indiscriminately over the upturned edges of the lower strata. The Oolites and the chalk were certainly once continuous all over this area, and extended far westwards; but they have been so denuded by running water, ice, and the influence of the weather, that their line of outcrop now lies to the east of Rutland; the post-tertiary beds being, in fact, to a large extent, made up of substances worn off or detached from the lower rocks.

Beginning with the oldest we will now examine the various stratified rocks of the county and endeavour to relate some facts of interest respecting them.

The *Lias* is said to have received its name from the manner in which quarrymen pronounced the word "layers," when describing the ribbon-like appearance that the alternate beds of limestones and shale present when a section is exposed in a railway-cutting or a brickyard. It usually gives rise to a heavy, wet, stiff clay district. It is, consequently, mostly pasture-land, for which it is well suited, being very rich in phosphates, and yielding a grass which is very fattening to cattle pastured on it. Extending westward into Leicestershire it constitutes the most famous hunting district in England.

The beds of the Lower Lias rarely appear on the surface in Rutland, for they are, to a certain extent, covered with drift, and are only to be seen in one or two places. In the north-west, near Whissendine, the strata are visible in a brickyard, and there are indications along the banks of the river Eye. They are blue clays with ironstone balls, and their total thickness is about 800 ft.

The Middle Lias or Marlstone (an impure limestone) is composed of two members which differ greatly from one another. The lower part is the Middle Lias clay, and the upper the Marlstone rock bed.

The Middle Lias clays are sandy at their base, but pass upward into beds of dark and light blue clays which are over 100 ft. thick in the north of Rutland, but diminish to less than half that amount as we go southward. These dark blue clays are full of fossils. Rounded nodules of impure limestone, called *septaria*, occur in layers. These *septaria* have usually selected some organic substance, like a shell, as a nucleus round which to collect, and when broken open frequently yield a rich harvest to the collector. Northwards, from Stockerstone, the Middle Lias clays occupy the bottom of the valley in which the Eye Brook runs, and an old brickyard at Belton was worked in them.

Above the Middle Lias clay is found a hard thick band of ferruginous limestone, called the Marlstone Rock. This often passes into good workable ironstone. Brown and friable when exposed to the influences of the atmosphere, when dug in deep pits or from under clay, it forms a tough, hard, crystalline rock of a blue or green colour. The upper beds are quarried for building-stone. Buildings such as the County Gaol at Oakham—no longer standing—were constructed of this material, and many fossils to be seen in the walls found vigorous employment for the schoolboy's knife. The lower bed being much harder rock is quarried for road metal. The superior hardness of the rock bed has withstood denudation better than the beds of clay above and below it, and in the north and west it forms a prominent feature in the landscape. About Teigh it stands out as a bold escarpment, abrupt to the west and gently sloping to the east. From this point it extends southward by Ashwell and Langham to Oakham, Egleton, and Lower Hambleton. The Marlstone forms the bottom of the Vale of Catmose, being 8 or 9 ft. thick, and yields, by its decomposition, the bright red-brown soil for which this part of the county is famous. It is highly productive and is almost everywhere ploughed and cropped, thus forming a very marked contrast to the adjacent slopes of Upper Lias clays which are used as pasture lands. The Marlstone can be seen standing out like a shelf along the sides of the valleys of the rivers Chater and Gwash; these streams having cut down through the drift and Upper Lias clays. Near Braunston, in the valley of the Gwash, the rock bed has been dug at several places, and is pretty well exposed. The commonest fossils are the oval-shaped brachiopod *Terebratula punctata*, and *Rhynchonella tetrahedra*, which has a plaited furrowed appearance. The inside of the shells, which usually occur in casts only, is filled with crystals of calc-spar.

Often bands occur, which quarrymen call "jacks," made up entirely of these fossils.

The Upper Lias clays have a total thickness of about 200 ft. Where they extend as a flat surface they are usually deeply covered with drift, washed off from the steep slopes which they commonly form. From Teigh the clay stretches southwards and forms the sides of the Vale of Catmose. It is then much broken up by outliers of Inferior Oolites, but extends without a break from Lyddington to Caldecott. At the base are to be found some thin layers of limestone, containing an abundance of *Ammonites communis*. The Upper Lias clays are worked in the brickyards at Pilton, where they furnish good material for bricks, tiles, and drain-pipes.

In Rutland there are two main divisions of the Oolitic rocks, the Inferior and the Great Oolite. The lowest beds are usually sandy, and as it is very prevalent in the county of Northampton, it has been named the Northampton Sand.

The main outcrop of this formation enters Rutland near Barrowden, in the valley of the Welland, and runs northward through Morcott to Luffenham, Whitwell, and Market Overton to Burley-on-the-Hill. Here it is about a mile wide, forming the beginning of the cliff which can be traced for 90 miles northwards into Lincolnshire. On its top runs the Roman Road known as "Ermine Street." The escarpment is very bold to the west, and all along the junction of the sandy beds with the impervious Lias clays beneath, springs gush from the hillsides.

South of Burley-on-the-Hill the river Gwash has cut down through the escarpment, and the tributary brooks which flow by the villages of Greetham and Exton have cut deep gorge-like valleys right through the Inferior Oolite limestones and sands down to the Upper Lias clay.

View from Wardley

About Uppingham the Northampton Sand occupies a considerable area, reaching to Stoke Dry, Seaton, Wardley, and Glaston. Here the flat-topped hills, so characteristic of this district, are seen to perfection. Between Ridlington and Preston the sand was evidently once worked for ironstone, for masses of furnace cinders are to be found. Small outliers are to be seen at Lyndon and Hambleton, and a tiny patch occupies the summit of Ranksborough Hill.

The Lincolnshire Limestone is a formation which has received its name owing to being most fully developed in Lincolnshire. The rocks are not everywhere exposed at the surface, large tracts being buried beneath accumulations of boulder-clay and gravels. In Rutland it is more exposed than in adjacent parts, and as it underlies the greater portion of the eastern part of the county, covering a triangular area, the three points of which are Thistleton, Stamford, and Barrowden, it is a prominent feature in the geology of the district. The Lincolnshire Limestone is extensively quarried at Ketton, Casterton, and Clipsham. That at Ketton is famous for its strength and durability, and has been used as a building-stone in many ancient and modern buildings in Cambridge, Bury St Edmunds, and Stamford, as well as the Cathedrals of Ely and Peterborough. It was the stone exclusively used in building the Law Courts in Fleet Street.

Reference may here be made to what is known as Collyweston Slate. At the base of the limestone, where the junction with the Northampton Sand occurs, are certain sandy beds, which, after being exposed to the action of frost, split into thin slabs that are excellent for roofing purposes. Sir Gilbert Scott and one or two other architects used them for church roofs. They are now only raised in the parish of Collyweston in Northamptonshire, and were it not that the fashion has been set in some small measure, in all probability that pit would be closed.

We now pass on to the Great Oolitic series. Referring to the table on page 24 it will be seen that this formation consists of three members, the first and third being clay, and the second limestone. The limestone in this series can be distinguished from that of the Inferior Oolite by the absence of the round particles, about the size of the eggs in the roe of a cod-fish, which has given a name to the whole of the Oolitic series. It is sometimes called Roe-stone. These little grains consist of carbonate of lime, arranged in successive concentric coats, round some particle of foreign matter which forms a nucleus.

The distinction between the Great Oolite limestone and Cornbrash—a local Wiltshire name for the soil cultivated in that county for the growth of corn, and used by geologists to indicate the strata which yield the soil—is not so easy. As a general rule the Cornbrash limestone is distinguished by its finer grain, its reddish-brown colour, and its peculiar wall-like bedding, as seen in weathered faces of rocks; while the Great Oolite limestone is coarser in grain, of a whitish colour, and weathers out in more solid blocks, with broad faces. The soil formed by the Cornbrash has usually a reddish hue, while the Great Oolite limestone has more commonly a black colour. A small portion only of this series comes to the surface in Rutland, but outliers form hills near Barrowden, Ketton, Casterton, Pickworth, and Clipsham.

The Upper Estuarine clay, which forms the lowest member of the Great Oolite series, rests directly on the Lincolnshire limestone. The clays are of various colours, usually blue, and make excellent firebricks. The beds form a cold, stiff, land and—even when well drained— an unkindly soil, giving rise to heathy tracts such as may be seen at Luffenham.

The Great Oolite limestones rest on the beds of clay previously described and form the low, flat-topped, hills already alluded to, which give so distinctive a character to the scenery in the neighbourhood of Little Casterton, Pickworth, and Stretton. The rocks give a good black soil, of high value to the farmer, and contain great numbers of small fossil oysters. Only very small tracts of the Great Oolite clays come within the confines of Rutland. At Barrowden and Clipsham there are outliers, while small patches remain near Essendine and about a mile west of Ryhall.

The Cornbrash, which is the last of the series, consists of a ferruginous limestone, containing a large number of fossils of the oyster kind. Only very small tracts are found in the county, about Belmisthorpe ; the main outcrop lying outside the boundary. There is, however, an outlier on Barrowden Hill, another north of Clipsham, and two small patches near Ryhall. The rock has no great economic value, as it is unfit for building-stone, but is extensively quarried for road metal. It appears to have been formed when the country underwent great subsidence. Towards the Fenland it passes beneath what is known as the Oxford Clay.

We have now come to what are known as the Glacial Deposits—the relics of the great Ice Age. It is believed that at a late period of geological history Britain must have been buried beneath a vast sheet of ice, similar to that which now covers Greenland. This ice played its part in rasping, grinding, and polishing the surface of the land, and the ice deposited a mass of debris, consisting of stony clays, gravels, loam, and sand. In some cases these deposits have given rise to lakes, and have frequently diverted and even reversed the courses of rivers.

There is no doubt that in Rutland, as in the greater part of England, the main features of the county, the hills, cliffs, plains, and some of the valleys, remain as they were before the Glacial period began, and it was over this surface that the Glacial deposits stretched in all directions, depositing boulder-clays, gravel, and sand.

This boulder-clay is of a blue colour, and filled full of pebbles and fragments of rocks, from the size of a marble to a cottage. It was evidently one of these huge masses which gave rise to the report, in the early days of geology, that chalk was to be found at Ridlington. The probability is that this mass, many yards in diameter, was conveyed by ice from the chalk escarpment which lies away to the north-east of the village.

On the sides of the valleys of the Chater, the Gwash, and the Glen, are found gravel-beds 30 or 40 feet above the present bed of the stream. These are Post-Glacial deposits and it is not unlikely that, if carefully searched, flint implements, bones of extinct quadrupeds, such as the mammoth, reindeer, etc., will be found in them, as is the case elsewhere.

When the ice melted and the face of the land returned to normal conditions, rivers began to course freely over the country. They cut new valleys in the Glacial deposits and ate their way into the Jurassic rocks beneath. By this means the drainage system which at present prevails in the county was evolved.

It only remains to mention the newest deposit in the county. On this subject Prof. Judd, who made an exhaustive geological survey of Rutland about the year 1875, and to whom we are indebted for much of the information contained in this section, says :—

"The flat bottoms of the valleys of the existing rivers are covered with a fine black loam or silt, which is still in process of formation, a constant accession to, and redistribution of, its material taking place as the result of ordinary river action. This loam is often crowded with the shells of those molluscs which live on terrestrial surfaces or in marshes ; and the deposit, which is in course of accumulation, in every respect resembles that found at higher levels in connexion with the old valley gravels. The flats formed by these alluvial deposits, which are very extensive (and are characterised by the valleys of the Welland and Gwash) are during the winter seasons for the most part under water : they are distinguished for their great fertility and constitute most admirable grazing lands."

7. Natural History.

In the previous chapter we learned something of the manner in which the rocks peculiar to Rutland were formed. There is unmistakable evidence that during the Carboniferous era, which produced the vegetation of which our coalfields are composed, practically the whole of England and Ireland was covered with water, with the exception of a strip of land extending from the Continent, westwards through the Midlands, to the coast of Ireland.

At one period the whole of the British Isles and a large portion of Europe was covered with a sheet of ice, which settled like a winding sheet upon it, and the previously-existing plants and animals disappeared.

The vegetation during the ice age was, doubtless, in many respects the same as that of Arctic Norway at the present day. The animals included the musk-ox, lemming, reindeer, and other still living arctic forms; but there were also some which have become extinct, such as the hairy mammoth and woolly rhinoceros. During the milder inter-glacial periods denizens of warmer regions found their way here; the hippopotamus wallowed in the rivers of the south of England, and the tiger, lion, bear, and other animals provided sport for our pre-historic ancestors, as their remains, found in caves and gravel-deposits in various parts of the country, testify.

By degrees, as the climatic conditions improved, the arctic vegetation was driven from central and western Europe, and plants loving a milder temperature, which

3—2

had doubtless been natives of Europe before the period of the Great Ice Age, once more occupied the places whence they had been driven. Arctic animals also retreated northward, or became wholly extinct. And thus, possibly as imperceptibly as it began, the long age of ice came to an end.

The animals and the plants, then, followed the receding ice, and were enabled to reoccupy the ground from which they had been driven, with the result that we have in England much the same fauna and flora as is to be found all over Europe. But the recolonisation was not complete before the land was once more changed. England, as we know it to-day, became separated from the Continent, and the Irish Sea now flows over the strip which joined her to the sister isle. Hence we have not as many kinds of wild animals in England as there are on the Continent of Europe, and Ireland has even fewer than there are in England. The same may be said of plant life.

The geographical distribution of animals and plants is a subject of the greatest interest. We can only touch the outskirts of it here, but one point of importance must be borne in mind, namely, that the local distribution of plants and animals undoubtedly depends to a great extent upon the natural features of the area in which they are found.

Although Rutland practically adjoins the Fen country, and is thought by many who have never visited the county to have a similar character, there are very few natural features common to the two districts. Only about 200 acres, or one five-hundredth of the surface of Rutland is water, and the nearest sea coast is 70 miles distant, so that our county offers but few attractions to the water and marsh-loving birds of the fenlands. The Burley-on-the-Hill and Exton Ponds are the only artificial waters of any size in the county, and the surface streams

Burley Ponds

are of small volume. But being a purely agricultural district and the population numbering only about one to five acres makes it, in some measure, favourable for certain forms of bird life.

Rutland seems to have been totally ignored by old writers on natural history. There does not appear to be a single instance in which allusion is made to it. The only information respecting animal life of by-gone times on record is that contained in a diary kept by Thomas Barker of Lyndon Hall, Rutland, who was brother-in-law to Gilbert White, the author of *The Natural History of Selborne*. This record[1] gives an uninterrupted series of observations for a period of 65 years, extending from 1736 to 1801, but they consist only of such entries as "First swallow seen." "First martin observed." There are, however, two interesting items, for they are initialled by the distinguished naturalist himself. "March 31st, 1736. A flock of wild geese flew N. G. W." "April 6th. The cuckoo heard. G. W." At this time Gilbert White was only about 16 years old.

Mr Haines, in his recently published book on *The Birds of Rutland*, estimates the bird population as follows :—Permanent residents 61, summer visitors 33, birds of regular passage 4, winter visitants 19, and occasional immigrants 20. The kite, buzzard, and raven, which were quite common a century ago, have disappeared. The peregrine and merlin are regular winter visitors, and the sparrow-hawk and kestrel, notwithstanding the efforts of gamekeepers to exterminate them, are still plentiful. The barn owl, too, is very common. Of game birds, the French partridge, which was introduced in 1850, thrives so well that on some estates its eggs are purposely destroyed. The quail is still met with now and then. Canada and Egyptian geese and the ordinary wild duck breed in

---

[1] The extracts are published *in extenso* in the *Rutland Magazine and County Historical Record*, Vol. III, p. 43, etc.

Burley and Exton ponds. Here the osprey is an occasional visitor and the heron a frequent one, though the latter does not nest in the county. The nightingale is abundant everywhere.

Speaking of the rare birds found in Rutland, Mr Haines says :—" The facts of more importance to British ornithology in general are :—The eighth instance of Bonaparte's Gull, included on the authority of an expert ; the unique nesting of the bee-eater ; the addition of Rutland to the counties where the pied flycatcher has been seen ; the recent appearance of the bearded tit in the county ; the late acquisition of the redshank as a nesting species."

The Bonaparte's gull was shot at Burley Ponds in 1897. A pair of bee-eaters, exceedingly scarce birds, built a nest in the bank of a pond close to Kelthorpe, a hamlet of Ketton, in the summer of 1868, the only instance on record for Britain. Male specimens of the pied flycatcher have been noted at Ridlington and in Exton Park. Two male bearded tits were shot at Burley Ponds January 18, 1905. The redshanks visited Rutland about 1890, and have since nested at Exton, Tixover, and other places along the Welland Valley, at Burley, and at Seaton.

There is nothing peculiar about the animal life in Rutland. In former days, when a large portion of the county was covered by dense forest, the larger and fiercer wild animals in all probability were plentiful.

Even the marten became extinct about 100 years ago ; and the polecat has been seen, trapped, or killed, in only five instances during the last sixty years. The fox, of course, exists to be hunted by one of the great packs of England—the Cottesmore—otherwise he would have been exterminated long ago. Stoats and weasels are common everywhere. Badgers breed at Exton and Burley, Preston, Stoke Dry, Thorpe-by-Water, and Wardley. They are much more plentiful than is generally supposed, and in the year 1900 no less than twelve were dug out at Burley. Otters are frequently to be found in the Welland, Gwash, and Chater, and are often killed by the foxhounds. They afford excellent sport for the Bucks Otter Hounds, which regularly hunt the district.

Hares are plentiful except in the southern part of Rutland and rabbits are to be found all over the county, those in the Teigh warrens being noted for their size and flavour. Squirrels are numerous in all the woodlands.

There are no wild deer in our county, but red deer were introduced into Exton Park about the year 1887 from Lowther Castle by the Earl of Gainsborough. The park is now stocked with a large herd of about 450 fallow deer, 52 red deer and 23 Japanese deer. Fallow deer have been there since the time of Charles I: there is also a large herd at Normanton, the seat of the Earl of Ancaster.

As there is so little water in Rutland there is not much scope for the fisherman, and the different specimens of fish present no marked features. Perch, pike, roach, dace, tench, carp, chub, and bream are to be found in most of the waters. Brown trout and the great North American lake trout in large numbers have from time to time been turned into the Gwash and Eye. In 1893, five thousand grayling were put into the Gwash, and both trout and grayling do well and provide excellent sport for the angler. The Gwash is considered one of the prettiest little trout and grayling rivers in England.

The grass-snake is common throughout the county, but the viper or adder has become very scarce latterly.

The limestone, mentioned in the Geological section as running from the coast of Yorkshire and passing through Rutland to the coast of Dorset, provides a home within the borders of the county for many flowers which could not otherwise exist on the heavy clay that forms so much of the surface. The conspicuous display made by some of the blue flowers is characteristic of the district. The bluebell or wild hyacinth (*Scilla nutans*) carpets most wonderfully some of the woods ; the deep azure hue gives a distinct colour to the landscape. Later in the summer the brighter blue of the flax (*Linum usitatissimum*) lights up many mossy dells ; then follows succory (*Cichorium intybus*). These two latter are only found in any quantity on the limestone. There are few bog plants in the county, the marshy land having been drained long ago. Up to three years ago the butterwort (*Pinguicula vulgaris*) had its home in a small bog, but recent drainage operations have brought about its destruction. That most beautiful flower the bog bean (*Menyanthes trifoliata*) flourishes in more than one spot, also the grass of Parnassus (*Parnassia palustris*) and the flowering rush (*Butomus umbellatus*); while the large bitter cress (*Cardamine amara*) brings Lincolnshire botanists to the borders of Rutland in search of it.

Hounds-tongue (*Cynoglossum officinale*), with its curious

smell, is found in all parts of the county. Henbane (*Hyoscyamus niger*), which used in former years to be cultivated for its medicinal properties, now crops up in most unexpected places only to disappear, perhaps, for some years before it puts in an appearance again. It is interesting to record the rediscovery (after many years'

Burley Avenue

disappearance) of the yellow star of Bethlehem (*Gagea lutea*) in a wood on the south-west side of the county. The mountain geranium (*Geranium pyrenaicum*) has occurred in a field below Catmose, Oakham, a very singular situation for such a plant. The woolly teasel (*Dipsacus pilosus*) is a pleasure to find. The narrow-leaved everlasting pea (*Lathyrus sylvestris*), the purple milk-vetch

(*Astragalus hypoglottis*), the autumn and field gentians, and the sulphur clover are found on the east, and the strawberry clover on the west side of the county. Rutland is rich in species of orchis ; the bee, the frog, the fragrant, the butterfly, the pyramidal, and the marsh orchis are all found within the borders of the county. Including grasses, 850 different species of wild flowers have been recorded ; and when it is remembered that Rutland contains only 97,273 acres, of which only 90 are heath, it will be acknowledged that the list is not a small one.

All kinds of forest trees grow well in Rutland. The oak must have been a familiar tree in ancient times in the Vale of Catmose, and doubtless the county town derives its name therefrom. Ash, oak, elm, beech, and horse-chestnut are plentiful ; and the magnificent avenues at Burley-on-the-Hill, Exton, and Normanton, show how suitable is the soil to the growth of fine timber.

8. Climate and Rainfall.

The climate of a country or district is, briefly, the average weather of that country or district, and it depends upon various factors, all mutually interacting, upon the latitude, the temperature, the direction and strength of the winds, the rainfall, the character of the soil, and the proximity of the district to the sea.

The differences in the climates of the world depend mainly upon latitude, but a scarcely less important factor is proximity to the sea. Along any great climatic zone there will be found variations in proportion to this proximity, the extremes being " continental " climates in the centres of continents far from the oceans, and " insular " climates in small tracts surrounded by sea. Continental climates show great differences in seasonal temperatures, the winters tending to be unusually cold and the summers unusually warm, while the climate of insular tracts is characterised by equableness and also by greater dampness. Great Britain possesses, by reason of its position, a temperate insular climate, but its average annual temperature is much higher than could be expected from its latitude. The prevalent south-westerly winds cause a movement of the surface-waters of the Atlantic towards our shores, and this warm-water current, which we know as the Gulf Stream, is the chief cause of the mildness of our winters.

Most of our weather comes to us from the Atlantic. It would be impossible here within the limits of a short chapter to discuss fully the causes which affect or control weather changes. It must suffice to say that the conditions are in the main either cyclonic or anticyclonic, which terms may be best explained, perhaps, by comparing the air currents to a stream of water. In a stream a chain of eddies may often be seen fringing the more steadily-moving central water. Regarding the general north-easterly moving air from the Atlantic as such a stream, a chain of eddies may be developed in a belt parallel with its general direction. This belt of eddies, or cyclones as they are termed, tends to shift its position, sometimes passing over our islands, sometimes to the north or south of them, and it is to this shifting that most of our weather changes are due. Cyclonic conditions are associated with a greater or less amount of atmospheric disturbance; anticyclonic with calms.

The prevalent Atlantic winds largely affect our island in another way, namely in its rainfall. The air, heavily laden with moisture from its passage over the ocean, meets with elevated land-tracts directly it reaches our shores—the moorland of Devon and Cornwall, the Welsh mountains, or the fells of Cumberland and Westmorland —and blowing up the rising land-surface, parts with this moisture as rain. To how great an extent this occurs is best seen by reference to the map of the annual rainfall of England on the next page, where it will at once be noticed that the heaviest fall is in the west, and that it decreases with remarkable regularity until the least fall is reached on our eastern shores. These western high-lands, therefore, may not inaptly be compared to an umbrella, sheltering the country further eastward from the rain.

The above causes, then, are those mainly concerned in influencing the weather, but there are other and more local factors which often affect greatly the climate of a place, such, for example, as configuration, position, and soil. The shelter of a range of hills, a southern aspect, a sandy soil, will thus produce conditions which may differ greatly from those of a place—perhaps at no great distance—situated on a wind-swept northern slope with a cold clay soil.

The Meteorological Office in London collects records

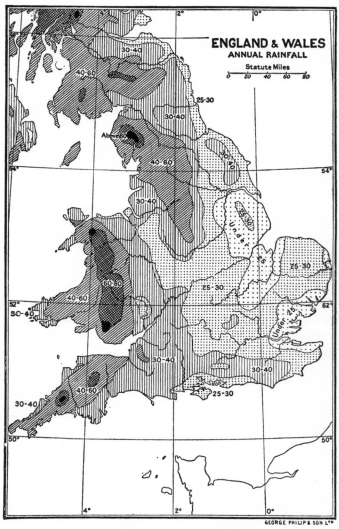

ENGLAND & WALES
ANNUAL RAINFALL

Statute Miles
0   20   40   60   80

25-30
30-40
40-60
Above 80
30-40
40-60
30-40
30-40
60-80
40-60
30-40
Under 25
25-30
25-30
25-30
Under 25
30-40
40-60
30-40
30-40
25-30

GEORGE PHILIP & SON LTD

(The figures give the approximate annual rainfall in inches.)

of temperature, rainfall, direction of the wind, hours of sunshine, etc., made by many observers stationed in various parts of the country ; and at the end of the year the averages or "means" are worked out for Great Britain as well as for counties and agricultural divisions.

The mean temperature at Oakham in 1910 was 48·8°. This differs by only 0·1° from that of the Agricultural Division that includes the counties of Nottingham, Leicester, Warwick, Northampton, Oxford, Buckingham, and Rutland. For the Division, it was 48·70°, for England 48·8°, and for the whole of Great Britain 48·1°. There were 1247 hours of bright sunshine, which was 78 hours less than the average. Comparing this with the North of Scotland Rutland got 88 hours less sunshine, while the mean temperature of North Scotland was only 45·7°. The highest mean temperature is held by the Agricultural Divisions which include (1) Cornwall, Devon, Somerset, and Dorset, (2) Kent, Surrey, Sussex, Berkshire, and Hampshire, which each registered 49·9°. Division (1) recorded 1559 hours of sunshine, while (2) obtained 1613 hours. The average sunshine for Great Britain was 1390 hours, for England 1432, and for Scotland 1295.

The highest temperature registered at Oakham in 1910 was 82° in the shade, and the lowest 19°. On August 10, 1911, the maximum thermometer recorded 92° Fahr. This was abnormal, as the mean maximum temperature for the week was only 80·7° in the shade. The mean range, in 1910, was 13·1°, which indicates a fairly equable climate. On only 32 days was there a range running from 23 to 31 degrees. The records

for the previous ten years show only one low reading.
On December 26, 1908, the thermometer began to fall
rapidly from 33°, and on December 31 reached 6°. Two
days afterwards it jumped up to 42°. During the ten
years 1901–1910, with the above-named exception, the
temperature never went below 13° Fahr. and touched
that point on only three occasions.

There are no less than 4538 persons in the British

The Welland in flood
(*showing a portion of the Seaton Viaduct*)

Isles who keep rainfall records. All these records are
dealt with in an annual called *British Rainfall*, so that
there is a ready means for ascertaining how much rain
has fallen in a given locality. Much more falls in the
West of England than in the East, as we have seen.
For example, in 1908 at Llyn Llydaw Copper Mill,
Carnarvonshire, the extraordinary quantity of 237 inches

of rain fell; while at Bourne, in Lincolnshire, only 15·63 inches were registered. An inch of rain equals 100 tons to the acre, or 64,000 tons per square mile. With a rainfall of 237 inches, therefore, covering an area of only one square mile, there would be more than 150 million tons of water to be collected, an amount which if stored in a reservoir 500 feet above sea level, would furnish power to drive an electrical plant sufficient to light a town as big as Liverpool throughout the year.

Taking a line directly from west to east (the wind blew in that direction 96 days in 1910) we find the rainfall as follows:

Montgomery	Salop	Stafford	Leicester	Rutland
61·3	35·51	32·09	28·99	24·6 inches

On 84 days the wind was south-west, and taking that line we have:

Glamorgan	Monmouth	Worcester	Warwick
55·73	50·28	31·9	29·89 inches

Leicester	Rutland
28·99	24·6 inches

We learn from this that there is a large diminution in the rainfall, especially on the direct line from west to east, immediately the coast county is left, and it steadily diminishes.

As previously stated, local factors, such as height above sea level, configuration, and position, materially affect local rainfall. Market Overton, standing on the

top of a bold escarpment 475 feet above the sea, registered
32·2 inches in 1910, while Lyddington, lying in a valley
152 feet above the sea, had only 23·26 inches. Again,
Oakham, 361 feet above the sea, nearly encircled by hills,
but open on the western side, registered 29·21 inches.
The number of rain days in Rutland in 1910 averaged
195, which does not compare unfavourably with that for
Great Britain, numbering 218. If, however, the various
stations where rainfall is measured be taken separately
it will be seen that there is great diversity owing to
varying local influences, e.g. Market Overton had 231,
Oakham 203, Ridlington 190, Whitwell 180, and Great
Casterton 171 rain days.

But the records for one year are, of course, not
sufficient from which to make general deductions.

Taking the records of Market Overton for 1910 we
have 23·2 inches. The average in Rutland for 1909
was 26·74, and in 1910, 24·6, while from 1875–1909 it
was 27·1 inches—a record which indicates very clearly
and forcibly the necessity for long-continued observation
before any reliable data can be established for such a
variable thing as mean rainfall.

9. People—Race, Dialect, Population.

It is very difficult to say what people were the earliest
inhabitants of Rutland. The study of Early Man, so
far as this county is concerned, has been almost entirely
neglected, and it is only since the institution of the Rutland

Archaeological Society, about ten years ago, that any progress has been made in this direction. It is possible that some evidence of the people of the Palaeolithic or Older Stone Age may be forthcoming, but, up to the present, Rutland has yielded no examples of their presence within its borders.

So far as we know, these first inhabitants of the British Isles knew nothing of the arts of the weaver, and can, therefore, have had no other clothing than the skins of beasts. They made rude flint implements by chipping, and used them as weapons of the chase. Although they made few advances in the homely industries of life, they had, curiously enough, a distinct feeling for graphic art, and on fragments of bone and ivory drew, with a sharp pointed instrument, representations of the creatures they hunted and used as food, many of which are long since extinct in the British Isles.

Afterwards came the Great Ice Age, to which we referred in a previous chapter, and Palaeolithic man ceased to exist in our country, and when it was repeopled it was by Neolithic man.

Of the presence of Neolithic man in Rutland we have evidence. Their civilisation was much higher than that of their predecessors, but they were often only nomad herdsmen, living in huts. Those who reached the Midlands may have lived in permanent villages, raising crops of oats or some rougher kind of grain for food. They were breeders of cattle, and knew how to weave cloth and to bake pottery. The ruder tribes, who subsisted entirely by their cattle, would naturally follow the herd, living through the summer in booths on the higher pasture grounds, and only returning to the valleys to find shelter from the winter storms. Strange to say they do not appear to have felt any of that passion for picture drawing which characterises their predecessors. In stature they never exceeded 5 ft. 9 in., and there is some reason to believe that they were a dark-complexioned race with black and curly hair.

There is a line of dry limestone downs running transversely from the Yorkshire Wolds through Rutland to the coast of Dorset. This is the region of the tumuli, and on its surface may be found the foundations of British huts. On the hills are their long boundary fences; below the edges of the hills rise innumerable bright streams, and by these springs, no doubt, were the settled habitations.

It is very interesting to note how the towns and villages are arranged along the outcrop of water-bearing strata, such as the junction of the Marlstone rock bed with the Lower Lias clays, and especially the Northampton Sand with the Upper Lias clays. Running the eye over the geological map, one would note at once the position of centres of population along such lines, which consequently determined the original settlement of the district. For to the early Celt, as to his successors, whether Saxon, Dane, or Norman, etc., an abundant supply of water was a paramount necessity.

Hitherto man had remained in ignorance of the metals, or rather of the use of them. But now he learnt how to smelt tin and copper, and their mixture, bronze, afforded him weapons and implements which advanced him greatly in civilisation. Nobody quite knows the origin of the invaders whose coming into this country marked the commencement of what is known as the Bronze Age. The tribes were known as Goidels or Gaels, a race of warriors and hunters who crossed the sea and established themselves in Britain, or as it was perhaps then called, Albion. In Rutland they left evidence of their presence at Cottesmore, where a hoard of bronze implements has recently been turned up in the ironstone workings.

In time man learnt how to work iron, and what we now term the Iron Age began. There is no sharp line of division between the two Ages, however, and no doubt implements of bronze—and even of stone—continued to be used.

The incursions of the Brythons (Britons) took place about 400 B.C. They were a Celtic-speaking race, and reached a comparatively high standard of civilisation in this district. The tribe of the Coritani, as the Romans called them, occupied part of Lincolnshire, the valley of the Trent, and probably Rutland, as this county was included in the division, when what is known as the Iceni Confederation was subdued by the Romans. The only probable evidence of the presence, in Rutland, of the men of this period is a find of two querns or grind-stones of the " bee-hive type," at Oakham and at Braunston.

The Roman occupation of Rutland, which possibly lasted about 300 years, had no great effect on the character of the inhabitants. That there was an extensive colony in the county is shown by the roads, camps, and buildings, and by the miscellaneous articles they left behind, of which more will be said in another chapter.

When the Romans left, about 410, the Saxons, Angles, and Jutes settled in various parts of England. There is

Quern of bee-hive type

ample evidence, from the remains found in recently-opened graves, that Rutland was the home of Anglo-Saxon settlers; but as there must have been extensive forests at that time all over the county, the population cannot have been dense at any point, and it is also probable that neither Saxons nor Anglians exclusively occupied Rutland.

Danish local names are unknown and Danish relics have never been recorded, hence it may be inferred that the Danes never had a settlement in the county. Nearly the whole of the names of the towns and villages are of Old English origin, that is to say, they are derived from the language of the people of the pre-Danish era, namely the Angles, or Saxons, or Jutes.

It is asserted, by some authorities, that the north-eastern part of Northamptonshire and Rutland are closely connected by dialect with Cambridgeshire. While the dialect noticed in the strip of country between Wisbech and Oakham may not be descended from the tongue of the Fenmen, it is considered more than probable that this area, with Cambridgeshire, formed a considerable portion of the territory in which a somewhat mysterious and isolated people called the Gyrwa, mentioned more than once by Bede in his *Ecclesiastical History*, so long maintained their independence and, no doubt, their peculiarities of language. Whether this is so or not, there certainly linger in some of the out-of-the-way villages in Rutland a number of provincialisms current in the counties of Huntingdon, Cambridge, and Lincoln. The following are a few of several hundred expressions recorded by the Rev. T. K. B. Nevinson, during a period of 16 years' residence in the village of Lyndon :—

" The wood is *doted* " (decayed).
" I've been *scuffling* " (hoeing).
" We've come a *booning* " (to boon = to help).
" We'd better *hock* it " (hock = rake (hay) into long rows).
" The hay's too *sammy* " (i.e. moist).

" The doves are very *cade* " (i.e. tame).

" Plants soon begin to *dowk* (droop) in dry weather."

" I love a *few* broth."

" My throat feels a bit *oasty*-like " (rough).

" When the grass is *frem* and growing."

" When the hay's *hessled* (dried) a bit."

" In a dry time the oats don't *stoven* out " (i.e. swell).

A hundred years ago the population of Rutland was 16,300 and a gradual increase took place until 1851, when it reached its highest point, namely, 22,983. Owing to the decay of agriculture during the last fifty years, the population steadily decreased until, at the census in 1901, it was only 19,709. This gives a density of 129 persons to the square mile, or an area of five acres for every man, woman and child. The last census of 1911 shows a very slight increase to 20,346. Westmorland, the most sparsely-populated county, has only 80 to the square mile, or one person for eight acres. In Lancashire there are 2554 persons to the square mile. Taking the whole of England and Wales there are 618 to the square mile ; and comparing Rutland with Middlesex, we see that it may be considered as one of the sparsely-populated counties, for in Middlesex there are 4855 persons to the square mile.

10. Agriculture—Cultivations, Stock.

Rutland is purely an agricultural county. The western portion is mainly grass and the eastern arable, the north and south being divided roughly between the two. Most of the grass is used for grazing purposes, though there is a fair amount of dairying. Grain was formerly grown over a much larger area than is now the case. Competition with other countries and our colonies, during the last thirty years, has caused agriculturists to turn their attention to raising cattle instead of growing corn ; hence during this period over 10,000 acres have been converted from arable to pasture land.

Rutland has an area, excluding water, of 97,087 acres. According to the report of the Board of Agriculture for 1911 the corn crops, consisting of wheat, barley, oats, rye, beans and peas, were grown on 19,015 acres, which represents about one-fifth of the area. Barley, wheat, and oats are the most important, 17,511 acres being given over to these. There is very little rye grown, only 22 acres being laid down, but 1482 acres are occupied by beans and peas. The Stamford district is one of the most famous barley-producing areas in England, its crops commanding a high price from the Burton brewers; and, in Rutland generally, barley has proved the most profitable of all crops, which accounts for the very unusual proportion of 9603 acres out of the 19,015 of corn crops being devoted to this produce.

The total produce in 1910 was 19,538 quarters of wheat, 47,471 of barley, 16,887 of oats, 2610 of beans, and 2167 of peas. Wheat yielded 36·32, barley 38·28, oats 40·49, beans 34·17, and peas 32·41 bushels per acre, which compares very favourably with the average yield for England. Rutland produces no great quantity of corn, being fortieth on the list for wheat. Lincolnshire takes first place with 775,894 quarters, and Westmorland the last with 437 quarters.

The green crops consist of potatoes, turnips and swedes, mangold, cabbage, kohl-rabi, rape, vetches or tares, and lucerne, etc., and they cover 6883 acres, or about one-sixteenth of the area of the county. Some 5255 acres are devoted to growing clover, sainfoin, and grasses. Part of this produce is for hay, and part not for hay, the land being broken up in rotation.

Only 48 acres are devoted to small fruit, strawberries, raspberries, currants, and gooseberries, and the area under orchards for growing apples, pears, cherries, and plums amounts to 86 acres only.

The largest proportion of agricultural land in Rutland is permanent pasture or grass not broken up in rotation. This area is no less than 54,686 acres or over one-half of the whole of the county. Land which does not produce any crops is said to be bare fallow, of which there are 1239 acres in Rutland.

The woodlands cover 3819 acres. Coppice woods, or those which are cut over periodically and reproduce themselves from shoots, represent one-quarter of the total area. About 147 acres of plantation are planted or replanted within a period of ten years ; and the remainder, amounting to 2669 acres, produce oak, beech, ash, chestnut, and hazel, the woods in several places being in their original forest state.

We now come to the stock that are reared in Rutland for various purposes. These include horses, cattle, sheep, and pigs. To say that in 1911 there were 20,420 cattle, and 90,076 sheep in Rutland would convey to the reader no idea of the important place the county holds for cattle and sheep-raising. For example, Leicestershire had 327,018 sheep or nearly four times the number raised in Rutland ; but the former county holds only the ninth place in the list, with 616 sheep per 1000 acres, while Rutland holds third place with 929 per 1000 acres. Kent

Pure-bred Shorthorn Bull

takes the premier position with 1010 sheep per 1000 acres, but Kent is a very long way down the list for the production of cattle, having only 98 per 1000 acres ; while Rutland holds the ninth place with 210. Leicestershire heads the cattle list with 286 per 1000 acres. It is a noteworthy fact, however, that, if we take the actual area in each county held in permanent grass, we find that

Rutland holds the first place for the production of cattle and sheep per acre. Kent comes second and Leicestershire holds the third place.

Rutland produces no great quantity of butter or cheese ; but Braunston, a small village on the borders of Leicestershire, is the home of the famous Stilton. The story goes that a farmer at Braunston sent, periodically, several cheeses to a friend of his who kept a coaching inn at Stilton, on the Great North Road, and travellers used to tell stories of the famous cheese obtained at lunch. Hence the name Stilton cheese, although not a single pound was ever made in the village which gave it the name.

The favourite cattle both for grazing and dairying are shorthorns, the Lincoln red being extremely popular. Lincolns are also the favourite sheep.

The breeding of horses is encouraged by the Rutland Agricultural Society, who offer premiums ; but Rutland has only 3691 horses, of which 2281 are used for agricultural purposes. This gives 38 per 1000 acres, while Cambridgeshire takes premier place with 62. There are 2620 pigs in Rutland, which gives about the same proportion per acre as horses. In this respect Suffolk easily takes first place with 190 per 1000 acres. The pigs chiefly bred are the large and middle white Yorkshires, or "Lincoln Curlycoats," with a few Berkshires here and there.

11. Industries and Manufactures.

The manufactures and industries of Rutland have usually been dismissed by topographers as " of no account." As stated in a previous chapter Rutland is purely an agricultural county. The climatic conditions would not allow, for example, the manufacture of cotton as in Lancashire, because a humid atmosphere is necessary for that industry, hence the county never has been and cannot become the home of any of those huge textile industries found in the North of England.

The old jury lists of the county, which contain the names and trades of persons summoned to serve on juries at Quarter Sessions indicate, to some extent, the trades in vogue 150 years ago. They present no unusual features except that flax-dressers are mentioned and there were weavers at Oakham, Uppingham, and Langham. In 1845 Uppingham and Oakham had between them eight straw-hat makers ; but the industry has now not a single representative. In the same year there were three chair-makers at Uppingham ; but these have followed the trencher-makers, who were so celebrated for their ware in medieval times, that their memory still lingers in the old proverb, " As round as an Uppingham trencher."

Before the introduction of railways and steamships, which have given facilities for the easy transit of agricultural produce, corn milling was evidently an important industry in the county. There are still windmills in working order at Whissendine and Morcott ; while at

Oakham, Manton, Ridlington, Cottesmore, Ketton, and other places are to be seen the dilapidated remains or sites of others. Water-mills are to be found in eleven villages, but with two exceptions all have been abandoned. A modern milling establishment at South Luffenham now represents this industry, where the latest type of machinery, driven by steam and electricity, has taken the place of wind and water.

A century ago, Oakham had "a considerable wool trade." The reeling of wool found occupation for numerous persons. The skilled workmen brought by Edward III into England from Flanders so materially improved the manufacture of cloth from the always excellent English wool, that the woollen trade became one of the most important industries in the country. The Statutes bristle with enactments for the protection and regulation of this manufacture, and, by a clause in an Act passed in the reign of Henry VIII, no person was allowed to exercise the handicraft of a wool-winder until he was furnished with a certificate from a wool-grower, and had taken an oath, before a Justice of the Peace, to abstain from fraudulent practices.

The extent to which this occupation was followed in Rutland may be estimated by the fact that in the year 1786 no less than fifty persons, residing in 28 villages in the county, were fined twenty shillings each at Quarter Sessions for false and short reeling. The oath they took "not to use deceit, craft, guile, or fraud in winding or folding wool" did not appear to lie heavily on their conscience. The introduction of machinery wiped the

village wool-winder out of existence.   Buyers from the
large Yorkshire manufacturing districts now visit farmers
and purchase direct, or the crop is sent to the annual wool
sale.

An interesting relic of the wool-trade is shown on p. 64.
Down to the time of George III, there were enactments
relating to the sale of wool, and officials called "tronators"
perambulated the county on horseback to receive the
custom or toll termed "tronage."   In almost every village
there was a pair of scales (wool beam) provided and kept
in order by the churchwardens.   The weight figured is
14 lb.   Wool is sold by the "tod" of 28 lb., and the
tronator would therefore carry two such weights slung
across his saddle by a leathern strap, the rectangular
perforations in the weight enabling him to do this.   The
weight bears the arms of George I, and is one of a pair
purchased at auction at a farm-house in Rutland a few
years ago, where they had evidently lain since the Act
was repealed.   Such weights are exceedingly scarce.

The brewing trade used to find employment for five
breweries, but at present the sole representative of this
industry is to be found at Langham.   A small boot
factory was established in Oakham a few years ago by
a Leicester firm, and gives employment to about 100
workpeople.

Several years ago the Midland Railway Company
built one of their huge forage stores at Oakham.   About
40 men are employed here, and the forage, gathered
principally from the fen district, consisting of hay, sainfoin, clover, oats, beans, maize, bran, and dried grain, is

Tod Weight

cleaned, crushed, mixed, and weighed in and out by electrically-driven automatic machinery. About 10,000 bags, each containing eight stone of provender, are turned out each week, from which between four and five thousand horses are fed.

12. Mines and Minerals.

There are traces at some spots in Rutland of the Marlstone rock having been dug and smelted for iron, but—possibly being found scarcely rich enough in ore to pay for working—the excavations were abandoned. These old workings show that formerly, when wood was plentiful, and charcoal consequently cheap, much ironstone—which is found in the Northampton Sand—was smelted in an imperfect fashion. Since the introduction of machinery for dealing with the problem of low-grade ores, it has been found that the ironstone in this county can be quarried and smelted at a profit. Instead, however, of attempting to smelt it on the spot, the mineral is sent by rail to the furnaces.

During the last twenty years ironstone quarrying has been going on at Cottesmore. The ore is worked by hand. The depth of the seam is about 9 feet, and this is overlaid with clay and sand to a depth of 5 or 6 feet. The output at Cottesmore is about 45,000 tons a year.

Quite recently a new area has been opened at Market Overton. Here the top soil, having a depth of about 6 feet, is removed by hand, the men being paid at so much per cubic yard. Steam diggers are then put to work on the ore, and scoop it up into railway waggons. At Market Overton two steam diggers are at work, one night and day, and the output is about 240,000 tons a year. Preparations have also been made for the working of the iron ore in the Uppingham district.

Ketton Freestone Quarry

Stone quarrying was carried on at an early date in the county. The most important are the celebrated Ketton freestone quarries, and those at Casterton and Clipsham. The Ketton quarries have been open about 400 years. The stone is a perfect oolite, free from shells. It is soft when quarried but hardens on exposure to the atmosphere, and is considered one of the best weather

stones found in the whole of England. It resists the
atmospheric influence of towns admirably. The stone
is found at a depth of from ten to forty feet from the
surface, which has to be removed, as it is not safe to
mine. The stone varies from 2 to 6 feet in thickness,
and blocks are obtained containing from one to one
hundred cubic feet; as at present quarried the grain is
rather coarser than that of the earlier output.

A considerable quantity of stone from the Ketton
quarries has been used from time to time to restore the
Tower of London. St Dunstan's Church in Fleet Street
and Baron Rothschild's tomb at Willesden Green Cemetery are two good examples of its use. The remarkable
Custom's House at King's Lynn is said to have been built
of Ketton stone, and was erected about 300 years ago.
A very large quantity has been used for church work
in Norfolk and Suffolk, and also in and around Leicestershire. It is being used in Exeter Cathedral, although
good local stone can be obtained and the freight to Exeter
is over one shilling per cubic foot; also at York Minster,
and for all the figure carving at Beverley. The artists
who handle it state that it retains its chisel life and carves
admirably.

In 1905 the quarries in Rutland produced 6315 tons
of limestone, sandstone and other material, and about
160 men were employed in dealing with it.

13. History of the County.

Little Rutland has been a good deal neglected by historians and chroniclers. No one has thought it worth while to make much of the doings of such a small patch of territory, which has never, perhaps, felt itself big enough to take any leading part in public affairs. Rutland did not rise to the dignity of a shire under the Saxon kings; and it was only after the Conquest that it was administered as a separate county. Few districts have so little to show in the way of changes and transformations. Trades and manufactures have nowhere established themselves. The population for some centuries was almost stationary, and there has been a distinct decline during the last fifty years.

The early history of our county can scarcely be separated from that of the surrounding districts. In British times it was part of a division of England inhabited by a tribe called by the Romans the Coritani, whose chief towns were Lincoln and Leicester. The extent of the territory occupied by this tribe is uncertain, but it probably included the counties of Northampton, Leicester, Derby, Rutland, and Lincoln.

The Roman occupation, under the Emperor Claudius, began in A.D. 43. After his legions became possessed of the greater part of the kingdom, the district subdued by Publius Ostorius, which included the country north of the Thames as far as the Humber and Mersey, was named Flavia Caesariensis, and Rutland was included in this division.

During the Saxon Heptarchy, when England was divided into seven parts, each of which had a separate ruler, it formed part of the kingdom of Mercia, which included the counties of Gloucester, Hereford, Chester, Stafford, Worcester, Oxford, Salop, Warwick, Derby, Bucks., Northants., Notts., Beds., Rutland, parts of Herts. and Huntingdon, and, after the union of all the kingdoms under one monarch, it appears to have been in the possession of the Crown.

The whole of Rutland—or, as it was then called, Roteland—was under the direct lordship of the Anglo-Saxon kings, and was generally assigned to the Queen for the time being. There is a curious survival of this state of things in the village of Ketton, which still pays an annual rent of a few shillings to the Crown, *pro ocreis Reginae*, to provide the Queen with leggings, an article of attire that may be represented by the high-buttoned boots of the present day.

Edward the Confessor, in his lavish affection for his favourite foundation of the Abbey of Westminster, bequeathed the whole of Rutland to the monks of Westminster. This grant, however, was not confirmed by William the Conqueror, except as to the tithes of the district. But the memory of the Confessor's gift is preserved in the connection still existing between the present Dean and Chapter of Westminster and the county. The town of Oakham is still divided into the separate jurisdiction of the "Lord's Hold" and the "Dean's Hold," and a triennial court is still held in the latter under the authority of the Dean of Westminster. The neighbouring manor of Barleythorpe has the unique distinction of having existed since the days of Edward the Confessor under the same lords, the Dean and Chapter being allowed as the virtual representatives of the Abbot and Brethren of Westminster.

The secular lords of Oakham have undergone greater vicissitudes and changes than the religious corporation. The Norman castle is said to have been built by Walcheline de Ferrers, of the great iron family, which bore three horseshoes as its heraldic device. These Norman "Iron Barons," who were Masters of the Horse of the Dukes of Normandy, were charged with the purveyance of horseshoes for the ducal cavalry, and might be considered as the feudal chieftains of the blacksmiths generally. Whether or not a remnant of their privileges is preserved in the curious horseshoe custom at Oakham, to which we shall more fully refer in a later chapter, is a question which will probably remain for ever unsolved.

The superstition that horseshoes affixed to a dwelling bring good luck to the owner derives no confirmation from the history of the owners of Oakham Castle. After the Ferrers family, the Mortimers held it—an unhappy, ill-fated race—and, succeeding these, the Bohuns. The tragic end of Edward, the last of the Bohuns, is familiar to us in Shakespeare's *Henry the Eighth*. Nor was the fate of Thomas Cromwell, who acquired the Manor and Castle of Oakham, among other spoil of the Bohuns, any more propitious ; while the Harringtons, who had flourished in the county for six centuries, came to a sudden end soon after acquiring the lordship of Oakham.

The short and baleful splendour of the two Villiers, Dukes of Buckingham, is hardly an exception to this record of ill luck. The spell was broken, however, by the Finches, whose fortunes were founded by Sir Heneage Finch, Recorder of London and Speaker of the House of Commons; while his son, Heneage the second, rose to be Lord Keeper, Lord Chamberlain, and Earl of Nottingham. Their descendants are still lords of the manor of Oakham.

Rutland did not rise to the dignity of a county until the reign of King John. Following the example of previous kings, he, in 1204, granted Rutland, Rockingham, and other places to Isabella, his queen, in dower, and in the grant Rutland is described as *Comitatus.*

Richard, Earl of Cornwall, King of the Romans, son of King John, was granted by his brother, Henry III, "the whole living of Rutland besides other land." On the outbreak of the Barons' War in 1264, he appears to have been the only adherent of the King in the county. Among the insurgent barons was Hugh le Despenser who held Ryhall. He was the last justiciary of England (Lord Chief Justice, *ex officio* Regent of the kingdom in the king's absence), took part in the fight at Lewes in 1264, and was killed at Evesham in 1265 with another prominent baronial leader, Peter de Montfort, kinsman of Simon, whose ancestors had held Preston in Rutland for several generations.

The history of the county during the fourteenth century appears to centre round the owners of Oakham Castle, and these troublous times were reflected in the

changing fortunes of this property. In 1309 Edward II
granted Oakham and the shrievalty of Rutland to his
favourite, Piers Gaveston. He came to a violent end
in 1312, being kidnapped by the Earl of Warwick and
executed on Blacklow Hill. Edward, Earl of Kent,
brother of Edward II, had the castle and manor conferred

Oakham Castle

upon him in 1321 and held them until his execution in
1330, when the estate reverted to the Crown. William
de Bohun, Earl of Northampton, was granted the property
by Edward III, but having no male heirs the estates again
reverted to the Crown; and Richard II, in 1385, bestowed
them on his favourite, Robert de Vere, Earl of Oxford.
He being banished the kingdom, Richard gave Oakham

and Rutland to his cousin Edward, with the title of the
Earl of Rutland, now used for the first time. He was
killed at the battle of Agincourt in 1415, and the estates
again passed to the Crown, being given by Henry V
to William de Bouchier and Anne his wife, who was
daughter and heir of Thomas, Duke of Gloucester, from
whom they passed to the Staffords and Dukes of Buckingham. From this time the holding of Oakham ceased to
have that importance in county affairs which had belonged
to it for over a century.

Rutland is especially connected with the stirring events
during the Wars of the Roses. Close to Exton, in the
small hamlet of Horn, ran the old British trackway
known in many parts of its course as Ermine Street,
but locally as Horn Lane. This lane was once the high
road to Lincoln, and to this circumstance Rutland owes
its possession of one historic, but little noticed, battlefield.
Here at a spot, still known as Bloody Oaks, on March 12,
1470, Edward IV encountered the hasty levies of the
adherents of the House of Warwick, chiefly men of
Lincoln led by Sir Robert Wells, better known as Lord
Willoughby ; and after a dreadful conflict in which
10,000 men are said to have been killed, the leaders
were made prisoners and their followers completely routed.
The King's artillery spread havoc and death among the
ranks to such an extent that the rout was turned into
panic and hasty flight, and the fugitives flung away their
tabards in order to run the faster. Such, at least, is the
generally-received explanation of the popular name of
the battle of Lose-coat Field.

Royal visits to the county were fairly frequent during the thirteenth and fourteenth centuries. It owed them

Jeffrey Hudson

to the fact that two main roads ran through the county, one, the before-mentioned Ermine Street, and the other

the road from Rockingham, through Uppingham and
Oakham, to Nottingham. Edward I, Edward II, Edward III, and Richard II, during their reigns, stayed in
various places in the county fifteen times.

Later came the visit of Queen Elizabeth, to which
we owe the popular story of the Oakham Horseshoe
Custom. James I visited the magnificent mansion at
Burley-on-the-Hill in 1621, where he was the guest of
George Villiers, Duke of Buckingham, and Ben Jonson's
masque of *The Metamorphosed Gipsies* was performed before him.

To Burley also came Charles I, and it was here that
the famous dwarf, Jeffrey Hudson, made his first appearance before the King and Court, served up in a cold pie,
from under the crust of which he leapt in the full attire
of a gallant page-of-honour. Evil days fell on the mansion
during the Civil Wars, when it was occupied by a Parliamentary force. These troops, finding the house scarcely
defensible, abandoned and set fire to it, and in the flames
disappeared all traces of the magnificence of the Villiers'
occupation. Only the stables remained, and the mansion
lay ruinous and deserted for many years, until Daniel
Finch, second Earl of Nottingham, bought it from the
second—the spendthrift—Duke of Buckingham. It was
rebuilt, and in the year 1908 was again attacked by fire,
due, in this instance, to the more prosaic cause of the
overheating of a chimney flue.

The Harringtons were an important family in the
county. Their chief seat was at Exton. Here, in
the church, is a fine monument to Sir James Harrington

and his wife, Dame Lucy, daughter of Sir William Sidney, and with this pair, by feminine descent, a large proportion of the more ancient nobility of the kingdom may claim to be connected. From this union it has been said that there have been descended, or nearly allied to their descendants, eight dukes, three marquises, seventy earls, nine counts, twenty-seven viscounts, and thirty-six barons, among whom were numbered sixteen Knights of the Garter.

The son of Sir James and Lucy Harrington was created Baron Exton in 1603 by King James I, and was the Lord Harrington of whom so much is heard in a subordinate way during the reign of that king. He is described as a bountiful housekeeper, and became the tutor and guardian of the Princess Elizabeth, daughter of James I, afterwards married to the King of Bohemia. It was, perhaps, in consideration of arrears due for the Princess's board and lodging, for which hard coin was difficult to extract from the wasteful, impecunious king, that Lord Harrington was granted the patent for coining farthings, a coin which hitherto had been supplied by private enterprise in the way of tradesmen's tokens, which were now pronounced illegal. The profit of coining these farthings, which were of much less intrinsic worth than the trifling value they represented, was no doubt a handsome perquisite ; though Lord Harrington did not live long enough to enjoy the profits or the popularity which attended the new coinage. But people called the new coins after him, and Ben Jonson in *The Devil is an Ass* writes, " I will not bate a Harrington o' the sum."

During the second Civil War, one of the first and greatest sufferers for the Royal cause was Henry Noel of North Luffenham. He was the younger son of Edward, Viscount Campden. Raids having been made by Lord Grey's forces in various parts of Rutland, Henry Noel thought it prudent to collect 15 or 16 of his neighbours to act as a guard at his house, at the same time officially informing one of the deputy-lieutenants of the county that it was "a measure of ordinary prudence, and he had no desire to molest the county or meddle with anie of their Armes." While this was going on at Luffenham, Lord Grey marched his troops to Brooke, hoping to capture Viscount Campden and his son Baptist Noel; but finding they had gone to Newark and had removed all the arms and ammunition in store, they appropriated some horses and goods and marched to North Luffenham, about 1300 strong. Noel's force had been increased by the accession of friends and neighbours to 200, about half of whom were armed with guns and the remainder with pikes and clubs.

On a demand to surrender, Noel answered that "he would stand on his defence while he had breath." A general skirmish ensued on both sides, and Catesby, a lieutenant of one of Lord Grey's troops, was killed, and the besiegers retired baffled in their effort to effect an entrance into the house.

Next morning the siege was renewed, some cannon were brought up and the house shot through, and the Parliamentarians set fire to the outhouses, barns, and stacks of corn. This so alarmed Noel's neighbours, for several of the cottagers were being burnt out of their homes, that in order to save further bloodshed he surrendered on condition that the fire should be extinguished. The house was given over to a general pillage. Noel was sent a prisoner to London, and after ineffectually petitioning Parliament for release and the restoration of his estates, which had been sequestrated, died in prison, from which the Lords gave permission for his body to be taken to Campden, in Gloucestershire, for burial.

Sir Everard Digby, one of the Gunpowder Plot conspirators, had an estate at Stoke Dry, a small village near Uppingham. To this fact may be attributed the interesting tradition that Stoke Dry was the place, and the parvise or chamber over the north porch of the church the room, in which the Gunpowder Plot was hatched. Like many cherished traditions tenaciously held for centuries, this one will not bear investigation. A very cursory review of the facts leads to the conclusion that there is no connection beyond the fact that Sir Everard Digby was born at Stoke Dry, and may have spent a few of his early years there.

14. Antiquities—Prehistoric, Roman, Anglo-Saxon.

While our knowledge of the history of the inhabitants of Rutland since the Roman occupation depends, to a great extent, on written records, any information about the inhabitants of earlier days is derived from an examination of the relics they left behind them. These earlier times are known as the "Prehistoric." They have been arranged for convenience' sake into the Palaeolithic or Old Stone Age, the Neolithic or Later Stone Age, the Bronze Age, and the Iron Age.

The rudely-chipped and unpolished stone implements characteristic of the Older Stone Age are invariably found in river-drift deposits or in caves. As Rutland possesses none of the latter and few of the former, it is probable that any finds of this Age will only be the result of a systematic search of the gravel deposits, but, up to the present, none have been recorded.

Between the Palaeolithic Age and the appearance of Neolithic man, a long interval of time elapsed, at all events in England. When the latter arrived they found the land virgin forest, bush, and bog. The new-comers were acquainted with many of the arts upon which civilisation was to be established. They were hunters and fishermen. They had begun to cultivate the earth and breed domestic cattle. They had even begun to make pottery. Their implements were still made of stone, but show a great improvement in their construction ; for many, unlike Palaeolithic implements, which are rudely formed by chipping, were ground down to a cutting edge and polished. These implements are not found in the gravel deposits, but are obtained from surface deposits or from barrows, their dead usually being buried accompanied by their stone weapons. A considerable number of these implements have been found in Rutland, consisting of scrapers and other worked flints, as well as arrow-heads. These have occurred at Ashwell, Brooke, Burley, Egleton, Empingham, Exton, Hambleton, Manton, Market Overton, Oakham, and Wing.

The most noteworthy discovery of Neolithic remains, however, took place in August, 1905, in a stone quarry at Great Casterton. Here was found a human skeleton accompanied by a ground and polished celt, a grinding-stone, and three flat fabricating stones for the manufacture of horn or bone-needles. The type of skull, together with the implements found near the skeleton, place it beyond doubt that they are of Neolithic origin.

The transition from the Stone to the Bronze Age is marked by a fresh race of men who invaded and conquered Neolithic people, and who brought a knowledge of bronze with them. They left traces of their occupation in most parts of the British Isles, but only one find has occurred in Rutland. This took place in the year 1906, in the ironstone workings at Cottesmore. It is apparently a founder's hoard, and consists of two socketed celts and the lower portion of two others, a spear-head, a narrow socketed chisel, three gouges, and a fragment of what has probably been a sword-blade. The presence of this latter relic suggests that the hoard belongs to the latter part of the Bronze Age, since swords are not found with remains attributed to the earlier part of this Age.

Succeeding the Bronze Age came a wave of immigration of a Celtic-speaking race called the Brythons (Britons), who introduced the Iron Age. Bronze was not entirely superseded by iron, but the latter metal entered largely into the manufactures. The only

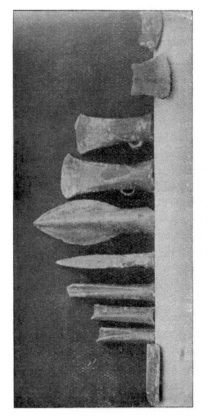

Bronze Age Hoard

(Found at Cottesmore in 1906)

remains of this Age found in Rutland are two querns or grindstones of the beehive type. These implements are associated with Iron Age remains; but as the type continued in use during the Roman occupation, it is questionable whether they can be definitely given a prehistoric date.

The Roman occupation under the Emperor Claudius began in A.D. 43; and by A.D. 47 the whole of the Eastern part of Britain up to the Humber, including the district now known as Rutland, was subjugated. Doubtless the Romans found the population scanty and peaceful, and following primitive agricultural pursuits. The only well-authenticated Roman road in the county is Ermine or Erming Street. It enters the county boundary at Great Casterton and leaves it at Thistleton Gap, a distance of about twelve miles. Another possible Roman road branches off from Ermine Street, about a mile and a half south of Stretton. It runs past Greetham and Thistleton, but all trace of it is lost after following the boundary line of Leicestershire and Lincolnshire nearly as far as Harston.

Near the place where Ermine Street enters the county, and also where it leaves, have been found earthworks described as Roman camps or entrenchments, and traces of habitations exist both at Great Casterton and Market Overton. It is probable, however, that the earthworks were of a non-military character, and possibly were not more than cattle shelters.

Few traces have been found of the existence of the villas of Roman landowners who may have settled in

the district; but tessellated pavements have been uncovered near Ketton, and it is possible that the finds
near Langham and Market Overton may indicate the

Roman Steelyard

presence of villas of a substantial character. The antiquities brought to light are of a varied character and
come principally from Market Overton and Thistleton,

where the ground has been searched. One of the most interesting objects, perhaps, is a *statera* or Roman steelyard. The sliding poise, usually the bust of an Emperor, is missing, as is also one of the hooks under the ball from which the article to be weighed was suspended. It is a double-action steelyard. The shank is graduated on each side. The steelyards manufactured to-day are on exactly the same principle.

A large number of Roman coins have been unearthed, possibly between two and three thousand, ranging from Claudius (A.D. 41) to Gratian (A.D. 383). Among the best examples (here illustrated) are a First Brass of Vespasian (1); a Second Brass of Trajan (2); three different First Brasses of Antoninus Pius (3 *a, b, c*); a Second Brass of Diocletian (4); three different coins of Carausius (5 *a, b, c*); a coin of Galerius Maximianus (6), and one of Magnentius (7).

The finds include an immense number of fragments of pottery of the plain and ornamental Gaulish red ware (commonly termed Samian) and specimens of local Castor ware with figures of animals. Among the former are pieces of bowls with mythological subjects, such as Hercules in the garden of the Hesperides, and Mercury, as well as animal and vine patterns.

The miscellaneous finds include querns, brooches, pins, rings, keys, *styli*, and other articles in bronze and bone; a short sword with curved bone handle, a knife with bone handle, parts of three steelyards, a handmill, iron clamps and hypocaust tiles.

Besides the finds at Market Overton and Thistleton

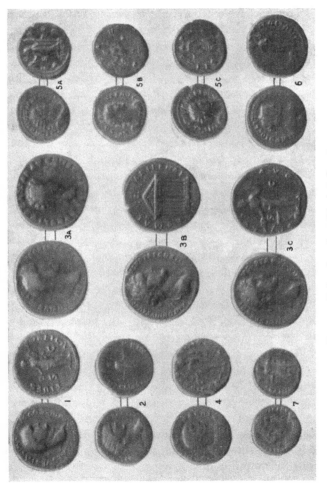

Roman Coins found at Market Overton

the following have come to light at various times in other parts of the county. A perfect brooch with pin, at Barrow, where pottery is also found scattered over the surface of the fields; coins in the vicinity of the Great Casterton Camp; pottery at Cottesmore; a coin of Antoninus in Exton churchyard; several square yards of tessellated pavement at Ketton; part of a fine bronze statuette of Jupiter or Neptune (now in the British Museum); a perfect bow-shaped bronze fibula at Seaton; and the base of a Roman column at Thistleton.

The area covered by the Northampton Sand mentioned in a previous chapter appears to be the resting place of Anglo-Saxon remains. Only two sites have, so far, been found in Rutland, namely, at Market Overton and North Luffenham. At the former place the finds have been due to the disturbance of the soil by ironstone digging; at the latter, graves were found quite near the surface where sand had been dug out for building purposes. In all probability further discoveries may at any time be made where agricultural operations and more especially ironstone workings are carried on.

The Saxon cemetery at North Luffenham was discovered so long ago as the year 1863, and the finds consist of spear-heads, knives, shield-bosses, buckets, urns, and a large number of bronze ornaments and glass beads. Among the more remarkable articles were two iron arrow-heads, as the bow was rarely used by pagan Anglo-Saxons. The fibulae or brooches found, while exhibiting a variety of type, were mostly of cruciform shape—a type peculiar to the Midlands.

The excavations for ironstone at Market Overton were begun in 1906 and as the surface soil to a depth

Anglo-Saxon Brooch found at Market Overton

of about 6 feet had to be removed in order to reach the ironstone, it is not surprising that some discoveries have been made. The finding of fragments of bone in urns points to the fact that an Anglo-Saxon cemetery was uncovered. Numerous examples of pottery, including an urn $10\frac{3}{4}$ inches high in complete preservation, were found. Of iron objects there were three spear-heads of different shapes, portions of shield-bosses, parts of sword-blades, a knife, and a pair of shears. Brooches ranging in size from two to five inches, as well as coloured beads, have been found in profusion.

The most unusual and interesting relic in the collection is a small saucer-shaped vessel, four inches in diameter, made of bronze, the bottom of which is pierced with a small hole. This is, no doubt, a water clock of a kind used by the early Britons. The bowl is placed on the surface of the water and allowed to fill through the perforation; on sinking in a definite time it is emptied and replaced. A series of experiments showed the average time taken to sink it was 62·9 minutes, hence it would answer well as a rough measurer of the hours.

15. Architecture — (a) Ecclesiastical. Churches and Religious Houses.

We will consider the architecture of the buildings in Rutland under three divisions, viz.: (a) Ecclesiastical, or buildings relating to the Church; (b) Military, or castles; (c) Domestic, or houses and cottages.

There is one fact worth noting with regard to these classes of buildings, and that is that—as indeed is universally the rule—the architecture of the county has been affected by the materials accessible. Thus we find that stone, wood, and bricks are used either according to the ease with which they can be obtained, or according to the wealth of the builders. As good building-stone is readily obtained in Rutland, this material largely prevails in the construction of churches and mansions.

The Ecclesiastical buildings in the county do not extend beyond churches—for there is no cathedral and no remains of abbeys, monasteries, or other religious houses within its borders. They are of various styles and of different ages, and have been classified as Anglo-Saxon, Norman, Early English, Decorated, and Perpendicular.

Forty-five out of its fifty churches are ancient, and there is quite an exceptionally large proportion of very fine and beautiful buildings. Oakham and Langham would rank in the first class in any district, while Ketton, Empingham, Exton, Cottesmore, Whissendine, Lyddington, Seaton, North Luffenham and Ryhall by no means exhaust the list of those considered admirable as whole buildings ; and those having interesting and beautiful details would include nearly every medieval church in the county.

Hardly any two churches are alike as whole buildings, nor do we find any two with details alike, although the same master hand may often be clearly traced. This is the more striking when Rutland churches are compared with those of other districts. In Norfolk or in Somersetshire we may see ten or a dozen churches, almost in adjoining parishes, having the same general form and outline, the same particular features and treatment. In

Rutland it is impossible to find any two churches which appear to have been turned out of the same mould, while as regards size, style, form, and treatment this difference is equally noticeable. Besides the unusually large churches at Oakham, Langham, Empingham and Lyddington, there are the unusually small ones at Pilton, Tickencote, Tixover and Essendine. The Gothic styles are all fairly represented, no one predominating, as is often seen in other districts. In respect of form, we find similar differences. Churches with two aisles, one, or none; with and without transepts; with and without chantry chapels. Some have bell-cots, some towers, some spires; and so we might run through a number of details. Probably nowhere in England, in so small an area, could such a great and remarkable variety be found.

A preliminary word on the various styles of English architecture is necessary before we consider the churches and other important buildings of our county.

Pre-Norman or, as it is usually, though with no great certainty termed, Saxon building in England, was the work of early craftsmen with an imperfect knowledge of stone construction, who commonly used rough rubble walls, no buttresses, small semicircular or triangular arches, and square towers with what is termed "long-and-short work" at the quoins or corners. It survives almost solely in portions of small churches.

The Norman Conquest started a widespread building of massive churches and castles in the continental style called Romanesque, which in England has got the name of "Norman." They had walls of great thickness, semicircular vaults, round-headed doors and windows, and massive square towers.

From 1150 to 1200 the building became lighter, the arches pointed, and there was perfected the science of vaulting, by which the weight is brought upon piers and buttresses. This method of building, the "Gothic," originated from the endeavour to cover the widest and loftiest areas with the greatest economy of stone. The first English Gothic, called "Early English," from about 1180 to 1250, is characterised by slender piers (commonly of marble), lofty pointed vaults, and long, narrow, lancet-headed windows. After 1250 the windows became broader, divided up, and ornamented by patterns of tracery, while in the vault the ribs were multiplied. The greatest elegance of English Gothic was reached from 1260 to 1290, at which date English sculpture was at its highest, and art in painting, coloured glass making, and general craftsmanship at its zenith.

About 1300 the structure of stone buildings began to be overlaid with ornament, the window tracery and vault ribs were of intricate patterns, the pinnacles and spires loaded with crocket and ornament. This latter style is known as "Decorated," and came to an end with the Black Death, which stopped all building for a time.

With the changed conditions of life the type of building changed. With curious uniformity and quickness the style called "Perpendicular"—which is unknown abroad—developed after 1360 in all parts of England and lasted with scarcely any change up to 1520. As its name implies, it is characterised by the perpendicular arrangement of the tracery and panels on walls and in windows, and it is also distinguished by the flattened arches and the square arrangement of the mouldings over them, by the elaborate vault-traceries (especially fan-vaulting), and by the use of flat roofs and towers without spires.

The medieval styles in England ended with the dissolution of the monasteries (1530–1540), for the Reformation checked the building of churches. There succeeded the building of manor houses, in which the style called "Tudor" arose—distinguished by flat-headed windows, level ceilings, and panelled rooms. The ornaments of classic style were introduced under the influences of Renaissance sculpture and distinguish the "Jacobean" style, so called after James I. About this time the professional architect arose. Hitherto, building had been entirely in the hands of the builder and the craftsman.

We may now turn to our county and note how far these various styles are represented in it. Of Saxon work there is but one example in Rutland. The tower arch of Market Overton exhibits the "long and short" work of the Anglo-Saxon builders, which may be clearly seen in the jambs. The imposts are rude and massive. The semicircular arch, though composed of several pieces, contains no keystone. What may have been two of the belfry "baluster" shafts are now built into the sides of the churchyard stile.

Let us turn now to the Norman style. There seems to have been a desire on the part of the architects who succeeded the Normans to preserve the doors of their predecessors and hence in many small churches, where

Market Overton Church

(*showing pre-Norman arch*)

all else has been swept away to make room for alterations, even in the Perpendicular style, the Norman door has been allowed to remain.   To this fact may be attributed the preservation, among many others, of the curious and well-known doorway at Essendine, the more curious and little-known doorway at Egleton with its elaborately carved tympanum, and the strangely-moulded doorway at Hambleton.

Tickencote Church: the Chancel Arch

As already stated, the semicircular arch is a distinctive mark of a Norman building.   There are, however, a few Norman arches of very curious shapes to be found in the country, approximating, in some cases, to that of a horse-shoe.   The wonderful quintuple arch at Tickencote is a famous specimen of this character, and, with the vaulted chancel of this church, finds a place in many

well-known books on architecture.   Other Norman
work to be found in the county includes the sturdy
tower at Tixover, the beautiful arcades at South Luffenham and Morcott, and the finely carved capitals at Seaton
and Stoke Dry.

We have said that no one of the four main Gothic
styles predominates in the county.   For the small number
of Rutland churches there is an exceptionally large proportion of Transition work, dating from the last quarter
of the twelfth century, which cannot be definitely classed
as Norman or as Early English.   There is probably no
district as good as Rutland for tracing the course of the
Transition from the Norman to Early English.   The
examples are many and widely distributed.

A local peculiarity is the late retention of the semicircular arch throughout the Early English style, in which
it is as common as the pointed arch.   Instances of this
may be seen at Manton and Great Casterton, in the
arcade on both sides of the nave ; in the south arcades
at Preston, Seaton, Edith Weston, and Clipsham, and in
the chancel arcade at Barrowden.   Many Early English
doorways, too, are round-headed, as at Whitwell, Barrowden, and elsewhere.

Good specimens of Early English work may be found
at Great Casterton and Empingham, the tower and spire
of Ryhall, the towers of Langham, Hambleton, and
Brooke, the beautiful arcades of Whissendine, and those
at Exton and Stretton, two very graceful capitals at
North Luffenham, and the exquisite belfry at Ketton.
The equal of this latter is not likely to be found in any

Langham Church

country parish in England.   The spire of Ketton, while very beautiful, is much too heavy for the light and airy

Oakham Church

belfry beneath, and this somewhat detracts from the composition.   It may be noted that the towers of Ketton,

Melton Mowbray, and St Mary's Stamford are very similar and are probably by the same hand. The spires of Ketton and St Mary's Stamford, although a century later in date and style, show an even greater similarity to each other.

We now come to the Decorated style. Of the early or Geometrical Decorated we have the towers and spires of Seaton, North Luffenham, and Cottesmore; all handsome, particularly the latter. Very little later is the steeple of Greetham, severely plain, but very noble and grandly proportioned, and the magnificent tower of Exton, built of the finest ashlar masonry, which, with its corner turrets and spire rising from an octagon, produces a most striking effect. Then there are the fine towers of Empingham and Oakham, and, perhaps, finer than any, the tower of Whissendine, where the belfry stage is most imposing, and has excellent details. The grand church at Langham is a good example of Decorated work; much of its exterior and nearly all its interior is in this style. More remarkable is the interior of Oakham, with its beautiful moulded arches, supported by clustered pillars having exquisitely carved capitals, all different, and of rare excellence. The number of good Decorated windows is large, some of the best and most highly ornamented are at Ashwell, and some of the most curious at North Luffenham.

Perpendicular work in Rutland is found in the nave, aisles, and clerestory at Lyddington. The nave arcades here are extremely fine and were probably built about 1490 by Bishop Russell of Lincoln. The north transept

Stretton Church

(showing bell-cot at the west end)

of Empingham, the chancel and aisles at Ryhall, many windows at Oakham and elsewhere, particularly the south transept windows at Langham and Whissendine (this latter now sadly mutilated) and the fine clerestory of the latter church, are all excellent specimens of the best Perpendicular work.

Bell-cots in place of steeples at the west end occur in some of the smaller Rutland churches. There are six altogether; Manton, Little Casterton, Whitwell, Essendine, Stretton, and Pilton; and it is worth remarking that all these are of early date, being either semi-Norman or Early English examples.

There is little to remark about the woodwork in the churches. Almost all the screens have disappeared; a fair one remains at Lyddington, where is also an interesting and rare example of the Puritan arrangement in the chancel, a communion table away from the east wall, with rails all round it. There is an extraordinary arrangement at Teigh, where the pulpit, reading desk, lectern and clerk's desk are all under the tower arch, a position possibly unique. The pulpit from which Jeremy Taylor preached, and which goes by his name, still remains at Uppingham, and there are some quaint Jacobean high pews at Brooke.

The fonts of the Rutland churches are usually plain, but there are interesting ones at Exton, Belton, and Market Overton, the latter with a very curious base, said to be an inverted Roman capital. In the one at Oakham, the base is probably that of a churchyard cross.

Of the religious houses none now remain. There

Teigh Church

was once a Priory at Brooke, founded by Hugh de
Ferrers, in the time of Richard I, as a cell to the
Priory at Kenilworth for Austin Canons, but beyond
the grass ridges indicating the position of the foundations,
all has disappeared. Hospitals were founded at Tolethorpe,
Great Casterton, and Oakham in the fourteenth century,
but with the exception of a charity for poor people, of
£14 a year, arising out of the latter, which serves to keep
the name alive, nothing is left except the chapel, used
as a Mission Church, the old bede houses having been
burned down many years ago.

16. Architecture—(*b*) Military. Castles, etc.

Castles, in the sense of fortified residences, were an
outgrowth or institution of feudalism. This was an
ancient system of government which prevailed in Europe
when lands were held by military services; those holding
land or buildings in feud, fee, or by tenure were bound
by an oath of fealty to serve the owner, at home or abroad,
in all wars and military expeditions when required.

In England there were few castles, properly speaking,
till the time of William the Conqueror, after which
a great many were constructed on the Norman model.
As Rutland does not possess a complete Norman castle
it would be well to give a brief description of one.

Approaching one of these old castles we should first
see a massive high wall, with an embattled parapet, towers,

and bastions.   This enclosed the courtyard, often of considerable extent, and was surrounded by a wide and deep moat.   The moat was crossed by a drawbridge, and a portcullis had to be raised before access was obtained. On one side of the court were the stables; in the centre was often a mound where the lord dispensed justice and where traitors and criminals were executed.   Another gateway led to the inner court, or bailey, and in this were the chapel, the barracks, and the donjon or keep, the strongest part of the castle, having walls of great thickness suited to form the last retreat of the garrison. The keep contained several rooms, one above another, and in the basement was a well to supply the garrison with water should they be surrounded by foes.

William I is credited with having erected 48 castles during his reign, but by the end of the reign of Stephen no less than 1115 were built, some hundreds of which were demolished by the succeeding monarch, Henry II.

Rutland appears to have had five castles or fortified enclosures, but four—Essendine, Woodhead, Beaumont Chase and Burley—have long disappeared, leaving nothing beyond the mounds, and, possibly, indications of the fosse surrounding the site they once occupied.

The Castle Hall at Oakham, now used as an Assize Court for civil and criminal business, is an excellent specimen of the type of fortified residence built in the latter part of the twelfth century.   There remains the walled enclosure, the fosse, now dry and covered with houses on the south side, and the banqueting hall, which is considered by competent authorities to be the finest specimen of domestic architecture of the twelfth century in this country.

There is a document of an interesting character, still preserved among the Inquisitions in the Public Record Office, which minutely details its precise condition in the 14th Edw. III (1340), and by this record we are enabled to trace out the site of some of the buildings within the enclosure which have since fallen down. "There is at Oakham," says the Inquisition, "a castle well walled, and in that castle there are one hall, four chambers, one kitchen, two stables, one grange for hay, one house for prisoners, one chamber for the porter, one drawbridge with iron chains, and the castle contains within its walls by estimation two acres of land; the aforesaid houses are worth nothing annually beyond reprises, and the same house is similarly called the Manor of Oakham. There is without the castle one garden, which is worth 8/- a year. Stews under the castle, with the fosse, the pasture of which is worth £6. 13s. 4d. a year. The park called the little park contains 40 acres, the herbage of which is worth £6 per annum, and the underwood 6/8. A windmill and a watermill are worth £8, and the presentation of the free chapel placed within the castle amounts to 100/-."

Of these buildings, there is nothing now preserved except the Banqueting Hall (see page 106). Its style is pure transitional, dating about 1160 to 1180, a period when the plain and massive Norman was merging into Early English.

It is smaller and earlier than the similar Hall at

Winchester, but in point of detail is considered more beautiful. This is particularly seen in the spirit and gracefulness of the different heads on the capitals, notably those of Henry II and his Queen, Eleanor of Guienne. The original roof was probably semicircular, like that still existing on the Bishop's Palace at Hereford. The oldest portion of the present roof consists of two beams put up by George Villiers, Duke of Buckingham, who also built the gateway.

No description of the castle could be written without something more than a passing allusion to the unique and remarkable custom for which the little county town of Oakham is celebrated. In the Hall hangs the curious and ever-increasing collection of horseshoes, numbering no less than 177 and ranging in size from that of the ordinary racer to the Brobdingnagian proportions of a shoe seven feet long.

The origin of the custom is lost in the mists of the past. Popular tradition dates it from the time when Queen Elizabeth passed through Oakham on her way to visit the great Lord Burleigh at Burghley by Stamford. The story goes that her horse cast a shoe in the street, and, in order to mark the event, the Queen there and then decreed that every royal personage or peer of the realm, on passing through Oakham for the first time, must give a horseshoe to the Lord of the Manor. On refusal the bailiff was to have power to take one by force from the horse's hoof. It is quite certain, however, that the custom is of much earlier date, for Camden mentions it as existing in his time (about 1586) and it is supposed

The Hall, Oakham Castle

to have come down from the Ferrers.   It has been said
that there is no other warrant for this conjecture than the
fanciful play on the word Ferrers, whose coat-of-arms was
semé of horseshoes.   Henry de Ferrers, who came over
with William the Conqueror, is thought to have received
that surname because he held the office of master of the
farriers or smiths in the invading army.

There is unmistakable evidence that Henry II (1161)
gave the manor of Oakham to Walchelin de Ferreris,
a descendant of the above-named Henry, and made him
Baron of Oakham.   He held it by tenure of service of
a knight's fee and a half, and it is to this Walchelin de
Ferreris that the erection of the castle and the exaction of
the horseshoe toll has been attributed.

As all sources of information have been searched
without result the writer puts forward what he considers
may be the solution.   The Parliamentary rolls at the
commencement of the reign of Henry I set forth a
petition of the Mayor of Dover that he may take toll of
every horse passing through the town to the amount
of a halfpenny, for the purpose of repairing the harbour.
There is no trace in the various records which have been
consulted that such an allowance was ever accorded to
Oakham, or to any of its proprietors.   The early existence
of the custom, however, seems to have established it as
a prescriptive right.   That Walchelin de Ferreris was
a truculent noble is evinced by the fact that in 1176
he was fined a hundred marks for trespassing in the King's
forests.   In 1181 he paid a hundred shillings for a pardon,
and in 1188 the sheriff returns him on the great Norman

roll as fined in one hundred pounds because of "a duel upon a robbery which was ill kept in his court." It requires very little stretch of the imagination to picture a Norman baron with such a record ordering his "seneschal" to claim a shoe from the horse of any other baron who rode through his territory ; and what probably began with a piece of ruffianism over seven hundred years ago has evolved into a curious custom which can be fully appreciated only by seeing the remarkable collection at Oakham.

The toll, of course, has long been commuted by payment for a shoe to be made of such size and design as suits the taste of the donor.

Speed, who visited Oakham about the year 1600, mentions the custom and also gives a short list of the shoes. Speaking with Lord Harrington he says : "that such homage was his due the said Lord himself told me, and at that instant a suit depended in law against the Earl of Lincoln, who refused to forfeit the penalty, or pay the fine." There is no evidence that Lord Harrington won his case as the shoe is not in the collection, but the great-grandson evidently did not dispute the custom, for a shoe was given in 1680 by Edward, Earl of Lincoln.

In this unique collection may be specially mentioned the horseshoe given by Queen Elizabeth, which is of iron, wrought in a curious chain pattern. It is not dated, but was probably given in the early part of her reign. The Prince Regent's horseshoe (George IV) is of solid bronze, stands seven feet high, and is said to have cost £50. Included in the collection are shoes contributed by the

Duke of York in 1778; Her late Majesty the Queen, when Princess Victoria, 1837; Queen Alexandra, when Princess of Wales, 1881; King Edward VII, when Prince of Wales, 1895; the Duke of Connaught, 1895.

We have here a curious custom which carries us back to a remote past and tells us something of the usages of our forefathers. There are but scanty materials from which a knowledge of such customs may be gathered and thence handed down for the instruction of posterity. In these modern days it is well to preserve as much as possible of the historic and picturesque past, so we trust the custom may long be continued.

17. Architecture—(c) Domestic. Famous Seats, Manor Houses, Cottages.

Under the heading of Castles we dealt with the oldest and most interesting specimen of domestic architecture in the county. For its small area Rutland is particularly rich in stone mansions and manor houses, the famous Ketton stone and that from Clipsham and other quarries in the neighbourhood being easily available.

Among the earlier buildings may be mentioned the old manor house at Ryhall, of thirteenth century date, which still retains its ancient cellar with a groined roof and other indisputable marks of antiquity. It was once the home of Hugh le Despenser, who was granted the manor by King Henry III. The house, or at least part of it, for it is now divided into two and has two owners, serves as a village alehouse.

Coming to the fourteenth century we have Tolethorpe Hall, a curious building which has been enlarged but still retains many of its original features. It is flanked on the south by two wings which, with the original building, form a courtyard. This was the home of the Burtons, an old Rutland family. Sir Thomas sold it in 1493 to William Browne, Merchant of the Staple, celebrated in our local annals as the builder of Browne's Hospital and the beautiful steeple of All Saints' Church, Stamford.

Next in date comes the picturesque old " Bedehouse " at Lyddington, formerly a palace of the Bishops of Lincoln. An oriel and three windows of an early Perpendicular character look out into the churchyard. The Hall is ceiled and decorated, probably by Bishop Longland between 1521 and 1547. The ceiling is in flat panels with a cornice of open carved work resembling fan tracery, unusual and very striking.

We must not forget to mention the old hall at Exton, with its grand gables, beautiful chimneys and ivy-clad walls, now fast falling into decay, which was built during the reign of Elizabeth ; and the fine old hall at Lower Hambleton, which, though now used as a farm house, is a good specimen of Renaissance work, with its interesting loggia, and old oak staircase, and panelling in the interior.

Burley-on-the-Hill is one of the most important mansions in the county and is certainly the most conspicuous. As mentioned in a previous chapter the old house was burned down by the Parliamentary army in 1645 and lay in ruins until Daniel Finch, Earl of

Exton Old Hall

Nottingham, bought it from the last Duke of Buckingham. Lord Nottingham rebuilt the mansion in its present form. It stands in a spacious, well-wooded park and on the south side commands a magnificent view of the adjacent country. The whole north front, extending one hundred and ninety-six feet, has a grand Doric colonnade, which consists of a series of single columns, thirty-two on each side. The large court is enclosed with iron palisades.

Burley-on-the-Hill

Normanton Hall, the seat of the Earl of Ancaster, was erected on the site of an ancient building belonging to the Mackworths by Sir Gilbert Heathcote, one of the first founders of the Bank of England and ancestor of the present owner. It is built of Ketton stone and although erected in the first half of the eighteenth century is very little weather dyed.   The style of architecture is Grecian of the Ionic order.

Passing from the larger houses let us glance at the buildings which add so much to the charm of our country-side.   In almost every village in the county, particularly at Preston and Braunston, may be found houses of local stone, with high-pitched gables and mullioned windows, built during the latter half of the seventeenth century.

Lyndon

These were the smaller manor houses, now converted into farm houses.   Despite many renovations these houses still retain their old character, and with their ivy-clad walls, roofs of grey Collyweston slate, and dormer windows, generally looking out on a garden full of flowering shrubs allowed to flourish at their own free will, show a picturesqueness which is a marked feature of the architecture of our county.

P. R.   8

It may safely be said that there is not a single village throughout Rutland which cannot produce examples of the handicraft of the thatcher. The reason for this is obvious. When the houses were built, reed or straw was the most easily obtainable material. The local Colly-weston slate, with which many of the larger houses were roofed, was expensive; and blue slate or tile quite out of the question owing to the absence of means of conveyance. Rutland appears to be the home of the thatched cottage. Even in its county town, in its High Street, are to be found houses which still retain it, and are from time to time re-roofed with this material. The walls are built of local stone and rubble, often three feet thick, which makes them cool on the hottest summer day and keeps out the cold in winter. There is no pretence to architectural style, but the artistic sense is quickened when they are seen nestling in some secluded lane, with their climbing roses and clinging ivy, and they certainly afford a pleasing contrast to the endless rows of "jerry-built" houses to be found in all the larger towns.

18. Communications—Past and Present. Roads, Canals, Railways.

In these days when a journey from Oakham to London, a distance of 96 miles, can be accomplished comfortably within two hours, it is almost impossible to realise the dangers and difficulties encountered by those, in days gone by, whose business required them to pass from one part of the country to another.

Before the Romans landed in England the only means of communication was by the British track-ways, such as the one known as Horn Lane, or paths trodden down through the forests.   The Romans, however, left behind them roads which were constructed with such extraordinary skill that even now many of the best highways in England

Section of Roman Road, near Great Casterton

are laid upon the ancient Roman foundations.   One of the most remarkable characteristics of a Roman road is the straightness of its course.   A good example is to be seen in the portion called Ermine Street, which traverses the eastern side of the county and is part of the Great North Road.   A section of the old Roman road has been exposed near Great Casterton.   The illustration above

8—2

will give some idea of the immense labour entailed in the construction of these roads.

The Anglo-Saxons adopted the roads and bridges left by the Romans in every part of the country. They gave the name of "street" (*straet*) to the roads, a word no doubt derived from the Latin word *strata* by which they had heard them designated by the Roman population. Such, doubtless, is the origin of the name of Stretton, formerly called Stretton-on-the-Street, a village which lies on the North Road about eight miles from Stamford.

Until the close of the seventeenth century few highways in England were more than mere open spaces, over which the public had the privilege of travelling. Goods and merchandise were conveyed by waggons, where the roads happened to be fairly firm, but more generally it required relays of pack-horses. All other travelling was performed on horseback. Ladies rode sometimes on side-saddles, but more commonly upon pillions seated behind their friends or their servant. In 1685 Lady Bridget Noel, who lived at Exton, made a visit to town. She arrived there " after a very troublesome journey," spending three nights on the road, at Buckden, Stevenage, and Chipping Barnet.

Arthur Young in his *Tours* describes the state of the roads general in all parts of the country at the end of the eighteenth century. He says : " Some of them had ruts four feet deep by measure, and into these ruts huge stones were dropped to enable waggons to pass at all ; and these, in their turn, broke their axles by the horrible jolting, so that within eighteen miles I saw three waggons lying in this condition." The mud in places was so deep and tenacious that in many instances it required the efforts of twenty or thirty horses to pull a waggon out of the slough in which it was immersed. If the state of the highways was bad, what must be said of the common roads? Not the slightest regard was shown for them. Whether they were mended or unmended mattered very little. Travellers were frequently obliged to dismount and lead their horses to avoid breaking their necks. Quagmires, sloughs, and bogs enveloped horses to the withers. Coaches were frequently overturned and this was often accompanied by loss of life. The streets in the towns were not much better, for in 1775 the inhabitants of Oakham were summoned to the Court of Quarter Sessions to answer an indictment respecting the state of the road at the Bargate. It was "in such decay for want of due reparation and amendment that the liege subjects of our Lord the King through the same way with their horses, coaches, carts, and carriages could not go, return, pass, ride, and labour without great danger to their lives and loss of their goods."

Richard Parkinson, who in 1808 wrote a survey of agriculture, remarks upon the roads in Rutland as follows: "The ruts are so deep that the traveller must keep the wheels of the vehicle in the ruts, by which both he and his horse are thrown and tost about in the most horrid manner imaginable; a chaise and pair has much greater difficulties to encounter, they go jostling one against the other, and keep slipping into the deep ruts, and are thus liable to fall every step they take, at the immenent risk of breaking the carriage, harness, etc."

The chief cause of the badness of the roads was the defective state of the laws. Every parish was bound to repair highways lying within its boundaries and the peasants were forced to give their gratuitous labour six days a year. If this was not sufficient hired labour was employed, paid for by a parochial rate. The injustice of this system will readily be seen. Such roads as the Great North Road passed through sparsely populated districts and, therefore, it is not surprising to find that parishes like Great Casterton and Stretton were absolutely powerless to keep the road in repair, worn as it was by the constant traffic between York and London.

Formerly, under the Turnpike Acts, many of the more important highways were placed under the management of Commissioners or trustees. The trustees were required and empowered to maintain, repair, and improve the roads committed to their charge, and the expenses of the trust were met by tolls levied on persons using the roads. But recent legislation has abolished the old toll-gate, and the last disappeared from Rutland some years ago. The Highways Act of 1835, and the various amending Acts passed since that date, have materially altered the state of the roads. In 1888 the entire maintenance of main roads was thrown upon County Councils and an Act of 1894 transferred to District Councils all the powers of Highway Authorities. It may be said that of all the proofs of social progress, which the country now exhibits to such a marvellous extent on every side, there is nothing more marked or wonderful than the rapidity with which we have improved and extended our internal communications.

Rutland has 320 miles of highways, most of which are in excellent condition. The cost of maintenance is a considerable item, running into some £10,000 a year. The two main roads passing through the county are the Great North Road and the road from Kettering to Nottingham. The former enters the county near Great Casterton, and, after following the line of the old Roman road (Ermine Street), leaves the county beyond Thistleton Gap, a distance of twelve miles. This was part of the old coaching road from London to York. There is a very large amount of traffic on the road and since the advent of the motor car the cost of maintenance has considerably increased, being at the present time £120 a mile.

The road from Kettering to Nottingham enters the county at Caldecott, and passing through Uppingham and Oakham, leaves it to the west of Whissendine. The gradients show the hilly character of the neighbourhood through which it passes, being from 1 in 10 to 1 in 17. This road traverses by far the most picturesque part of the county.

The railway systems which supply Rutland are the Midland, the London and North Western, and the Great Northern. The Midland line from St Pancras to Nottingham enters the county by the Seaton Viaduct over the Welland, a most remarkable structure, with its 82 arches, each of 40 feet span. This viaduct is

The Great North Road at Tickencote

70 feet high and was completed in 1878 at a cost of £82,000. The line runs through Manton and Oakham and leaves the county about two miles beyond Whissendine. The Midland Company has also a branch line from Peterborough through Stamford. It enters the county between the latter town and Ketton, and having a station at Luffenham, joins the main line at Manton. This

The Seaton Viaduct over the Welland

provides a connection with the Great Eastern and Great Northern systems.

The London and North Western, coming from Market Harborough, enters the county at the south-west corner, near Caldecott. Proceeding to Seaton it sends a branch to Uppingham, and, dividing its main line, passes through Morcott and Luffenham, and thence to Stamford and Peterborough over the Midland Company's line, the other branch leaving the county near Barrowden with a station at Wakerley.

The Great Northern crosses the extreme south-east part of the county with a branch line having stations at Essendine and Ryhall, and its terminus at Stamford.

Of Canal navigation in Rutland there is none. In 1793 there was a canal opened which extended from Oakham to Melton where it joined the Wreak and Eye navigation which connected it with the Soar and Trent. This canal has been disused for a number of years, and the part lying between Oakham and Ashwell is now preserved by the Oakham Angling Society.

19. Administration and Divisions—Past and Present.

When the Danes came pouring into England in the reigns of Egbert, Ethelwulf, and Alfred they found the country full of unwalled *tuns* or villages and scattered country houses called *burhs*, which were protected only by a moat and stockade. Of these the invaders soon made short work, and immediately started to secure themselves by erecting strong fortifications. The hardest part of King Alfred's work was the capturing of these Danish strongholds, but, having done so, he imitated his opponents by rearing many *burhs* which he filled with armed men. The *burhs*, or boroughs as we now call them, became the homes of loyal Englishmen, keen to resist an invading foe and also keen for commercial enterprise. Markets were here fixed, and in order to guard against theft it was ordained that purchases and sales should take place within their limits.

The union of a number of *tuns* for the purpose of judicial administration, peace and defence, formed what is known as the *Hundred* or *Wapentake*. These terms are both found in Anglo-Saxon records, but how they acquired their geographical application is a vexed question. Perhaps the simplest theory is that, in England, both names belonged primarily to the popular Court of Justice, and secondarily to the district which looked to the Court as the judicial centre.

Wapentake is a word said to be derived from that clashing together of their weapons whereby the Scandinavians signified their assent to propositions laid before them by the masters of their assemblies. But the term, used for a geographical division, is found only in Anglian districts—in Yorkshire, Lincolnshire, Nottinghamshire, Derbyshire, Northamptonshire, Rutland, and Leicestershire—and is probably a relic of Scandinavian occupation. The term hundred, like wapentake, first appears in the laws of Edgar as a name for an English institution, but has its origin far back in German antiquity. It has been regarded as denoting a district occupied by a hundred warriors with their families. This, probably, accounts for the fact that in some places the hundred has a large area, the population being scanty ; in others, a much smaller area, the population being denser ; hence it cannot be referred to any system of land measurement such as the *hide*. At the time of the Domesday Survey there were in Rutland (or so much of it as bore that name) two wapentakes. That of Alfnodeston contained two hundreds, one half being in Turgastune wapentake and the other in Brochelston, in Nottinghamshire. The other was the Martinsley wapentake and contained one hundred.

The *tuns* or townships had a chief officer, called the Reeve, who settled small disputes, but important matters were taken to the Hundred Court, which was held once a month. The presiding officer was called the *Ealdor*, a title from which our present word Alderman is borrowed. This court was attended by the lords of lands within the hundred, or their stewards representing them, by the parish priest, the reeve, and four best men of the township. The Judges of the Court were the whole body of freeholders, but, as much inconvenience would arise from uncertainty of qualification or attendance, a representative body of twelve men was chosen as a judicial committee of the Court. Here we have the origin of our present day Juries.

In the year 1315 it was found, by inquisition, that Rutland was divided into four hundreds. Three of these were held by the King, who received the profits. The fourth was also in the King's hands owing to Thomas, son and heir of Guy de Beauchamp, Earl of Warwick, not being of age.

At the present time Rutland is divided into five hundreds, Alstoe, Martinsley, Oakham-Soke, Wrangdike, and East Hundred.

Before the passing of the Act of 1848 for regulating

County Constabulary, each hundred had two chief constables, who, according to the Statute of Winchester in 1285, were chosen "to make the view of armour, suits of towns and of highways, and such as lodge strangers in uplandish towns, for whom they will not appear."

An ancient court which still exists in Rutland, though shorn of its powers, is the Court Leet (i.e. the Judicial Court of the Leet or hundred), in which was administered the "view of Frank-pledge" and the loyal oath of all who had attained the age of fourteen. This remarkable Saxon institution contributed very effectively to the prevention of crime as well as to the detection of offenders. Should anyone take a stranger in, and allow him to stay three nights under his roof, and the stranger afterwards committed any crime, the person so harbouring him was considered as having made a *pledge* for him as for one of his own family ; and if the offender absconded, was compelled to make amends to the injured person.

The chief officers of the county are the Lord Lieutenant and the High Sheriff. The office of Lord Lieutenant was first instituted in the year 1553, when "Commissions of Array" (issued to three or four of the principal men in the county, for the purpose of collecting, organising, and disciplining an effective military force) were superseded by "Commissions of Lieutenancy." From that time until 1872 the Lord Lieutenant was *ex-officio* the head of the Militia forces in the county. Since 1872 his chief duty has been to recommend to the

Lord Chancellor suitable persons to be made County Magistrates.

Until the year 1660 Rutland had not a separate Lord Lieutenant, that officer being appointed for the combined counties of Leicester and Rutland. In the year 1648 a petition of the nobility and gentry of Rutland was sent to the King. The petition set forth that "although Rutland had been time out of mind an entire county of itself, with a Sheriff, Assizes, Justices of the Peace, Knights of the Shire, and other incidents thereto as an entire county, yet in military affairs it had been heretofore subject to the Lord Lieutenant of another county." The petitioners prayed his Majesty to appoint a Lord Lieutenant for Rutland alone and in the year 1660 Baptist, 2nd Viscount Campden, was appointed for Rutland. The Lord Lieutenant holds his office for life and is also Custos Rotulorum, that is, the officer entrusted with the custody of the rolls and records belonging to the Sessions of the Peace. Both appointments are honorary.

The High Sheriff is the chief officer of the Crown in every county, who does the Sovereign's business; the Crown, by Letters Patent, committing the custody of the county to him alone. The office is an ancient one, dating back before the Norman Conquest. In early times practically the whole administrative system of the State, as it affected local divisions, was worked through the Sheriff. At the present time his ordinary duties, such as the execution of writs, he delegates to his deputy, the Under Sheriff. The Sheriff performs in person only such duties as are either purely honorary—for instance, attendance upon the Judges on circuit—or such as are of some dignity, and public importance, as the presiding over elections and the holding of County meetings, which he may call at any time.   He is appointed annually by the Crown.   Rutland's record of Sheriffs dates from the year 1129.

For the administration of Justice, Assizes are held twice a year at Oakham Castle, when both civil and criminal business (if any) is transacted.   The Judge has, however, seldom anything to do beyond congratulating the Grand Jury on the absence of crime, and accepting from the High Sheriff a pair of white gloves.

Quarter Sessions are held four times a year in Oakham, the President being the Chairman of the Magistrates.   Petty Sessions are held monthly.

Formerly nearly the whole of the business of the county was conducted by the Magistrates, or Justices of the Peace, at Quarter Sessions, but since the passing of the Local Government Act of 1888 most of this has been transferred to the County Council.   This body meets once a quarter at Oakham and is composed of seven Aldermen and twenty-one Councillors.   The latter are elected triennially ; but the former are co-opted, or selected by the Council itself, either from its own body or otherwise, and hold office for six years.

The duties of the County Council include rating and assessment ; the management of pauper lunatic asylums (Rutland in this combines with Leicestershire) ; the maintenance of main roads and bridges ; the registration and polling of parliamentary electors ; the appointment and control of the police, which is carried out in conjunction with the magistrates through a Standing Joint Committee ; and generally the carrying out the various Acts of Parliament passed from time to time. The expenditure for police, main roads, elementary and higher education and other charges, amounts to over £20,000 a year.

Rutland is subdivided into three Rural Districts, namely, Oakham, Uppingham, and Ketton. The Rural District Councils carry out the provisions of the Public Health Acts. Oakham has recently (1911) become an Urban district and so manages its own affairs. Parish Councils come next. These superseded the old Vestries which used to elect the churchwardens and the highway surveyor. But the Act of 1894 changed this, and each civil parish, if the population exceeds 300, must now have a parish council to conduct its business. There are 58 civil parishes in Rutland.

For Poor Law administration the county is divided into two Unions, namely Oakham and Uppingham, the members of the District Councils in those towns acting as the Board of Guardians generally.

Ecclesiastically, Rutland is in the diocese of Peterborough. Until 1541 it was part of the diocese of Lincoln and in the Archdeaconry of Northampton ; but, on the institution of the bishopric of Peterborough, the Archdeaconry was given to that see, and Rutland is now a Rural Deanery, divided into three parts, under the Archdeaconry of Oakham. There are 42 benefices. Five of these, until recently, were what are termed

"Peculiars," being Prebends belonging to Lincoln. They were Ketton *cum* Tixover, Lyddington *cum* Caldecott, and Empingham.

The Parliamentary History of the county began in 1295, when Simon de Bokmister and Robert de Felix-thorpe were returned to the "Model Parliament." Although the smallest county in England, Rutland continued to send two members to Parliament for nearly 600 years. After escaping the Reform Acts of 1832 and 1867, it fell a victim to the third Reform Act of 1885, and since that year has been allowed to return one member only.

20. The Roll of Honour.

Small though it is, Rutland can boast of numerous worthies who were either natives, or made the county their home. In a work of this character it would be impossible to mention in detail all those whose names may fairly claim a place on its Roll of Honour ; we shall, therefore, have to be content with a somewhat indiscriminate selection. The fact that parts of the county were constantly reverting to the Crown, and as constantly being given by the reigning monarch as rewards for services, brings the names of great families prominently before us. With these we briefly dealt in our Historical section and other chapters. Perhaps one of the earliest worthies to whom Rutland may lay claim is William the Lyon, who became King of Scotland in 1165. He held the manor of Exton. The property remained in the hands of his descendants for nearly two centuries, including that of Robert de Brus, or Bruce, called the "Competitor," from his having claimed the Crown of Scotland on his descent through the female line from William the Lyon.

The connection of Elizabeth, Queen of Bohemia, daughter of James I, with Exton attracts our interest from the many romantic traditions which surround her story. Soldiers bled for her whom they called "The Queen of Hearts," and men of letters were no less devoted to her service. Lord Orford allows her a niche among his "Royal Authors," while Donne, Daniel, and other poets sang her praises. She was consigned to the exclusive care of Lord Harrington and his lady, to be educated and maintained, and spent ten years of her life at Exton, where her memory still lingers in the fine avenue known as the Queen of Bohemia's Ride.

A character who figured much in the pages of English History during the first half of the fifteenth century was Humphrey, Duke of Gloucester, younger son of Henry IV, who claimed the regency. He is supposed to have been the founder of the Bodleian Library and, under his patronage, which he readily extended to men of letters, many learned foreigners were induced to settle in England, bringing with them the arts and learning of the east and south. He held the manor of Stretton.

The county has given the titles of Earl and Duke to several families, some of royal origin. Edward, eldest son of Edmund of Langley, fifth son of Edward III, was created Earl of Rutland in 1389. He it was of whom it is said that he behaved most gallantly at the Battle of Agincourt, "but being corpulent, through much heat and thronging, was smothered to death." Edmund Plantagenet, third son of Richard, Duke of York, was created Earl of Rutland in 1445, but was assassinated by Lord Clifford, after the Battle of Wakefield, in 1460. Thomas Manners, son of George, Lord Ross, received the title in 1528. In 1703 the then Earl was created Marquis of Granby and Duke of Rutland, and the title continues to the present day with the owner of the historic castle of Belvoir in the neighbouring county of Leicester.

We turn now to statesmen and lawyers. The Finch family, which has long been associated with Rutland, seems to have had an hereditary eminence in the study and practice of the law, ever since the reign of James I, in whose time Sir Henry Finch was a learned sergeant-at-law and author of *Nomotechnia*, a valuable work, though, perhaps, partaking somewhat of the quibbling spirit of the period. Sir John Finch was Lord Keeper of the Seals in 1617 and Speaker of the House of Commons in 1628, famous as the Speaker who was held down in the chair to prevent him adjourning the House. Sir Heneage Finch was Recorder of London in 1620 and Speaker in 1626. Heneage Finch, 1st Earl of Nottingham, was Attorney-General 1670, Lord Keeper 1673, Lord Chancellor 1674, and created Earl of Nottingham 1681. Burnet, the only author who has made an adverse censure on him, says "he was a man of probity, and well versed in the laws." His second son, Heneage, 1st Earl of Aylesford, was Solicitor-General from 1679 to 1686. He was called "silver-tongued Finch," and was leading counsel at the trial of the Seven Bishops. Daniel Finch, 2nd Earl of Nottingham and 7th Earl of Winchilsea, became First Lord of the Admiralty in 1681, was twice Secretary of State, and became Lord President of the Council in 1714. From this office he was dismissed, in 1716, by George I, for advocating leniency to the Jacobite peers. Macaulay says that he was the only honest man of his day. His son, Daniel, the 8th Earl, was returned at the head of the poll as M.P. for the county, in the first political contest ever held in Rutland. In 1741 he was made Lord Commissioner of the Admiralty and created a K.C.B. Our Parliamentary annals carry the connection of the Finch family almost down to the present day ; the last representative being the late George Henry Finch, who represented Rutland for an unbroken period of forty years —a somewhat unusual record—and who, before his death, became the " Father of the House of Commons."

Rutland has given to the Church one Archbishop of Canterbury. Simon de Langham, born in the village of that name, was chosen to the See in 1366. Like many others who have held the title he previously had experience in affairs of State, for he was appointed Treasurer of England in 1360 and Lord Chancellor in 1363. He went to the Papal Court at Avignon, and it was while returning to England that he died. His body was interred in the church of the Carthusians at Avignon, but three years later the remains were conveyed to the Abbey Church of Westminster, in which great convent he had been successively monk, prior, and abbot, and where, in the Chapel of St Benedict, his gray marble tomb, with effigy of alabaster—the oldest and most remarkable ecclesiastical monument in the abbey—remains to this day.

Jeremy Taylor, the well-known divine, and author of *Holy Living and Holy Dying*, and other works, was Rector of Uppingham from 1638 to 1642. He was called upon to attend Charles I in his capacity as Chaplain at Oxford. He subsequently accompanied his Royal Master through much of the Civil War and was presented by him shortly before his execution with his watch and some jewels. He was taken prisoner in the Royalist defeat before Cardigan Castle in 1645, but shortly afterwards was released; and, his living at Uppingham having been sequestered, supported himself and his family for some time by keeping a school in Carmarthenshire, where some of his best literary work was done. He was made Bishop of Down and Connor at the Restoration, in 1661, and died in 1667.

Though not a native of this county, but of Leicestershire, the Rev. William Whiston, philosophical writer, but better known as the translator of the works of Josephus, passed many of his later years at the house of his son-in-law, Thomas Barker of Lyndon. His body lies buried in Lyndon churchyard.

Of historians Rutland can claim only two. James Wright, antiquary and miscellaneous writer, whose *History of Rutland*, published in 1684, forms the basis for every subsequent work dealing with the history of the county, was the son of Abraham Wright, Vicar of Oakham 1660-90, himself the author of several religious and other works, including *Delitiae Delitiarum* (a collection of epigrams) 1637 and *Parnassus Biceps* (a collection of poetical prose) 1656.　James was a

Jeremy Taylor

barrister of the Middle Temple and among other works published *Historia Histrionica*, 1699, and *Country Conversations*, 1694.　Our other local historian was Thomas Blore who, after making extensive collections relating to the topography and antiquities of Hertfordshire, which formed the nucleus of Clutterbuck's history of that county, settled at Manton in our county. Here he began illus-

Jeremy Taylor's Pulpit

trating, in various ways, Wright's *History of Rutland*, adding genealogies, emblazoning arms, at which he was very expert, and in other ways making it a splendid volume. The book is now in the possession of a local resident. From this beginning may be attributed, in great measure, the history of the county which he afterwards wrote and partly published in 1811. The work was of such merit that it became the model on which all subsequent works of the same kind were formed, including Surtees's *Durham*, Clutterbuck's *Hertfordshire*, Baker's *Northamptonshire*, etc. But death put an end to his labours before the second part was published, and the manuscript, though known to have been sold by a London bookseller in 1887, cannot now be traced. Blore was for a brief period editor of *Drakard's Stamford News* and died in London, November 10, 1818, aged 53.

Among miscellaneous writers must be mentioned the Rev. Edward Bradley ("Cuthbert Bede"), author of *The Adventures of Mr Verdant Green*, who was rector of Stretton 1871–83. The greatest difficulty was experienced in finding a publisher for the book, but subsequently, when the three parts, issued in paper covers, were bound in one volume, over 100,000 copies were sold. Bradley was a contributor to many well-known magazines, and his separate publications extended to over 20 volumes. During the years 1852 to 1889 he made no less than 1157 contributions to *Notes and Queries*. He died in 1889 and was buried at Stretton.

Vincent Wing, who was born at North Luffenham in 1619, claims a place on our roll chiefly as a noted astrologer, astronomer, mathematician, and almanac-maker. In his time, not only the common people, but members of the highest ranks of society, had a passion for penetrating into the future. It is not, therefore, surprising that Wing's forecasts of the "aspects" and "influences" of the planets, which formed the subject-matter of his almanacs, received general favour. The Stationers' Company published the book and considered a sale of 50,000 a year an indifferent one. He published *Computatio Catholica*, a dictionary of the most remarkable accidents, occurrences, etc., beginning with the creation and extending to the year 1665. He was a prolific writer. His chief and most useful work appeared in 1652, entitled *Astronomia Britannica*. This was a complete system of astronomy on Copernican principles and included numerous and diligently-compiled sets of tables. He died in 1668 and was buried at North Luffenham. Tycho Wing, a noted astrologer and philosopher, whose portrait may be seen in the Hall of the Stationers' Company in London, was grandson of Vincent's nephew John. He lived at Pickworth. In addition to being coroner of Rutland from 1727 to 1742 he edited *Olympia Domata*, the almanac founded by his great-great-uncle, from 1739 until his death in 1750.

Rutland can boast of one local poet. John Banton of Teigh, son of a labourer, received his education at a village school, where, he declared, he was never taught a grammatical lesson. A collection of poems from his pen appeared under the title of *The Village Wreath*, followed by *Excursions of Fancy*, in 1824, consisting of pastoral, descriptive, and other poems. Ten years later he published *The Sulliot Chief*, a dramatic poem, founded on an attempt made by the notorious Ali Pasha to subjugate the Suliotes, an independent tribe of Greeks; an account of which the author found in the *Monthly Magazine*. He was schoolmaster at Teigh for a number of years and died in 1848. His verses indicate the possession of a vivid imagination, an excellent knowledge of the classics, and facility of expression.

Rutland has given three Lord Mayors to London. Sir John Brown, who was Lord Mayor of London in the reign of Edward IV (1481), was son of John Brown of Oakham, and his son, Sir William Brown, was Lord Mayor in the reign of the two succeeding Henrys. Sir Gilbert Heathcote, who purchased the Normanton estates in 1729, was one of the projectors of the Bank of England, and an Alderman, Lord Mayor, and M.P. for London.

We conclude our roll of famous men with a most remarkable character, the celebrated dwarf, Jeffrey Hudson, who was born of normal parents at Oakham in 1619, his father being a drover in the service of George, Duke of Buckingham. He was well-formed and good-looking, although only 18 inches in height (see p. 74). He remained at this height until his thirtieth year, after which he grew again, but never exceeded 3 feet 6 inches. At a banquet prepared in honour of the King and Queen, Jeffrey was served up in a cold pie for the amusement of the Court. He was taken into the service of Queen Henrietta and knighted by the King. Some time afterwards, when returning from the Court of France, where he had been sent to bring over a French servant, the vessel in which he sailed was boarded by a pirate and he was carried prisoner to Dunkirk, where he fought the famous battle with a turkey-cock which Sir William Davenant, the Poet Laureate of the day, ridiculed in a mock-heroic poem entitled *Jeffreidos*. Some time after his liberation he was sent on another foreign mission. On this occasion he was taken captive by a Turkish pirate, carried to Barbary, and sold as a slave. Having been ransomed, he returned once more to his native shores to find the country engaged in civil war. Jeffrey immediately took up the King's cause and was made a captain of horse in the royal army. He took part in a charge with Prince Rupert on a troop of Roundheads, near Newbury, where both the prince and the tiny knight had to beat a hasty retreat.

Queen Henrietta took the dwarf with her when she left England for the city of Nevers. It was here that the famous duel was fought between Jeffrey and Mr Crofts, a member of the Queen's household. Crofts looked upon the affair as a joke and took nothing with him but a large squirt filled with water, with which he intended to put out both his small opponent and the priming of the pistol. Jeffrey, who was mounted on a horse, eluded the shower of water aimed at him and killed Crofts with a shot from his pistol. He was forced to fly, but returned to England, where he lived in obscurity for some years. Being suspected of complicity in the Popish Plot he was thrown into prison and, although released shortly afterwards, died in 1682. His little waistcoat of blue satin, slashed and ornamented with pinked white silk, and his breeches and stockings, in one piece of blue satin, may still be seen in the Ashmolean Museum at Oxford.

21. THE CHIEF TOWNS AND VILLAGES OF RUTLAND.

(The figures in brackets after each name give the population in 1911, and those at the end of each section are references to the pages in the text.)

Ashwell (246), a small village three miles north-east of Oakham. The church is an ancient fabric, restored in 1851. Externally it is of Decorated character, with the exception of the outer doorway, which is Early English. There are three ancient monuments of extreme interest. An altar tomb supports a large marble slab with incised effigies of John Vernan, and Rose, his wife, about 1479. Their son, John, was prebendary of the Cathedral Churches of Salisbury and Hereford. In the north chantry is an alabaster effigy of a priest, almost life size, in full sacerdotal robes, c. 1485. In the south chantry is a hollow wooden effigy of a knight, the date of which may be about 1280. (pp. 27, 80, 98, 122.)

Ayston (75), a village one mile north-west of Uppingham. The church is in the Perpendicular style. In the churchyard is a curious monument representing two sisters who had only two arms between them. They were, however, so well able to employ themselves in spinning that they earned a sufficient sum to purchase a field, which they left for the benefit of the poor of Uppingham parish.

Barleythorpe (261), a hamlet one mile north-west of Oakham. The Hall is a stone mansion in the Elizabethan style, but modern. (p. 70.)

Barrow (110), a small village and chapelry in the parish of Cottesmore. A portion of an ancient cross stands near the centre. (p. 86.)

Barrowden (460), a village and parish picturesquely situated on the north bank of the river Welland. There is an interesting church in the transitional Norman, Early English, and later styles. The tower has a broach spire. The ancient custom of " Rush bearing " is still kept up by strewing the porch floor on the dedication festival. (pp. 18, 28, 30, 31, 32, 95, 122.)

Belmisthorpe, a hamlet prettily situated on the east bank of the River Gwash in Ryhall parish, with which the population is included. It was originally the property of the famous Lady Godiva. (pp. 20, 32.)

Belton (297). This village is on the borders of Leicestershire about three-and-a-half miles from Uppingham. It stands on a height on the north bank of the Eye Brook and within the old bounds of Leighfield Forest. This is a place of historic interest. In the reign of Edward II the manor was held by the Blounts, afterwards Lords Mountjoy. It passed from them to the Hazlewoods—the effigies of Sir Thomas and Clemence, his wife, dated 1539, are in the church—and from them to the Verneys. The Old Hall of the Verneys is an extensive building which still presents many peculiar features of interest, although it is converted into a farmhouse. The church is an ancient one, containing transitional Norman, Early English, and Decorated work, and has an embattled western tower with numerous carved gargoyles. (pp. 18, 26, 100.)

Bisbrook (190) is pleasantly situated on high ground, one mile east of Uppingham, the scenery being very picturesque.

The village is surrounded by hills and dales and the whole is given up to orchards and gardens producing a large amount of fruit and vegetables. The church is modern.

Braunston (357) lies in a sequestered valley, through which the river Gwash runs, near Leighfield Forest on the borders of Leicestershire. It is an old-fashioned village with very lofty gabled stone houses. The church is of the thirteenth century and has a fine late transitional Norman doorway, semicircular headed, and with a row of bold dog-tooth ornament in the hoodmould. There are some brasses to the Cheseldyn family dated 1596 and 1642. Much Stilton cheese is made in the village. (pp. 19, 27, 53, 60, 113.)

Brooke (80), is situated two-and-a-quarter miles south-west from Oakham on the bank of the river Gwash. Some fine transitional Norman work is to be found in the church, and a most interesting Norman doorway. There is also a monument with the effigy of Charles Noel, who died in 1619, whose father, Andrew Noel, was for many years high sheriff and knight of the shire. A little to the north of the village is the site of Brooke Priory, of which only a few mounds now remain. Of a post-Reformation residence built on the same spot only an Italian arch, of the same date as portions of the church, and a pigeon house now exist. (pp. 8, 19, 77, 80, 95, 100, 102.)

Burley-on-the-Hill (203), situated two miles north-east of Oakham, commands most magnificent views over the Vale of Catmose into Leicestershire. The house, one of the finest seats in the kingdom, is described elsewhere. The park is enclosed by a stone wall nearly six miles in circuit. It contains 1085 acres, in many parts covered with very large oak, elm, beech and other forest trees.

There is nothing striking about the church, except the lofty Decorated fourteenth century tower, but inside will be found the recumbent effigies of a knight in armour and a lady, probably fifteenth century work.   There is also a beautiful marble monument by Chantrey.   (pp. 9, 19, 28, 29, 37, 39, 40, 43, 75, 80, 103, 110.)

Caldecott (270).   This village stands on the north side of the river Eye, near its confluence with the Welland, at the southwestern extremity of the county.   It is ecclesiastically connected with the manor of Lyddington.   There is an ancient church here, originally built about the thirteenth century.   The spire was shattered by lightning in 1798 and rebuilt with Weldon stone, an inferior kind of freestone.   (pp. 12, 18, 28, 119, 121, 129.)

Casterton (Great), or **Casterton Magna**, or **Bridge Casterton** (288), is situated on the Great North Road, on the north side of the river Gwash, two miles from Stamford.   The Roman Road called Ermine Street passed through the village and no doubt the Romans had a station here, for there is a camp on the eastern side of the village.   The church is an ancient structure in the Early English style and still retains all its original windows, including those of the clerestory, which are circular. Under a flat arch, on the outside of the south aisle, is the recumbent effigy of a priest, in excellent preservation, though it is apparently of fourteenth century date.   (pp. 20, 30, 31, 50, 66, 80, 82, 86, 95, 102, 115, 118, 119.)

Casterton (Little) (189), lies in a hollow on the south side of the river Gwash two-and-a-quarter miles north-west of Stamford.   The church is an ancient one, consisting of nave with clerestory, aisles, chancel, porch and double bell-turret at the west end.   In the floor of the chancel is one of the finest engraved medieval monumental brasses in England, containing representations of Sir Thomas Burton and his wife, in the costume of the latter part of the fourteenth century.   The knight is in chain mail, and wears the collar of SS (the livery of the Lancastrian Kings).   The church contains a number of other interesting monuments.   (pp. 20, 32, 100.)

Clipsham (134) lies near the borders of Lincolnshire, nine miles north-north-west of Stamford. The church is an ancient fabric consisting of nave, with aisles, chancel, south porch tower and spire, and shows Norman, Early English, and Decorated work. The east window is filled with beautiful stained glass by Wailes, representing the Crucifixion. The church was restored in 1853. (pp. 30, 31, 32, 66, 95, 109.)

Cottesmore

Cottesmore (459) is pleasantly situated on the road between Oakham and Greetham, about two miles from Ashwell station on the Midland Railway. The oldest portion of the church is Norman or twelfth century work; the inner doorway is very rich with zigzag ornament. (pp. 14, 53, 62, 65, 80, 86, 89, 98.)

Edith Weston (268) lies six miles north-east of Uppingham. The village belonged to Editha, queen of Edward the Confessor and daughter of Earl Godwin. It was probably the most western place which belonged to her in Rutland, hence its name. The church is an ancient fabric with a tower, crowned by a light spire, and much admired by architects. (pp. 5, 95.)

Egleton (120) is a small village about a mile from Oakham on the south-east side. There was a guild here at an early period formed for the maintenance of a priest to sing mass " for ever." The church is a chapelry of Oakham. It has a range of Norman arches on its northern side and there is a curiously carved tympanum under a Norman arch in the south doorway. (pp. 27, 80, 94.)

Empingham (639) is a large village situated on the north side of the river Gwash, six miles east of Oakham. It was anciently a market town. The church is a fine specimen of Early English architecture, with a handsome tower, surmounted by a short crocketed spire. The nave arches on the south side are Norman and on the north Early English. Most of the windows are lancet-shaped and some are very rich in tracery. In the north wall of the north transept is a tomb attributed to one of the Normanvilles. (pp. 17, 19, 80, 89, 90, 95, 98, 100, 129.)

Essendine (215) is a small village near the river Glen on the borders of Lincolnshire. A battle is said to have been fought here between the Saxons and Danes. There was once a castle here, probably occupied as late as the reign of Queen Elizabeth by a branch of the Cecil family, but nothing now remains except a portion of the moat. The church is a small structure with a double bell-cot. The south doorway is Saxon work. In the tympanum is a carved figure of the Saviour with his hand resting on a book and an angel on each side. (pp. 11, 32, 90, 94, 100, 103, 122.)

Exton (613), a large village, situated in a pleasant open valley five miles from Oakham. The church is one of the finest in the county. The lower part of the tower is square with turrets and pinnacles at the corners, and above it rises an octagonal tower, from which springs a lofty taper spire.

There is here an interesting series of very fine monuments connected with the families of the Lords of the Manor. In the chancel on the north side is a monument to the memory of James Harrington, Kt, and his lady Lucy, both of whom died in 1591. On the opposite side is an excellent example of sculpture by Nollekens. It is in white marble, and in memory of Baptist Noel, 4th Earl of Gainsborough, who died in 1571. In the north transept is a magnificent marble monument to the memory of Baptist Noel, Viscount Campden, who died in 1683, an exquisite specimen of the art of Grinling Gibbons. At the west end of the church is another monument by Nollekens, in memory of Lieut.-General Noel, who died in 1766, and there are other tombs and monuments of interest.

The New Hall, which is about one hundred and fifty yards from the site of the old family residence, the south wing of which was destroyed by fire in 1810, is a large mansion in the Elizabethan style built at various periods since 1811. (pp. 9, 29, 37, 39, 40, 43, 73, 75, 80, 86, 89, 93, 98, 100, 110, 116, 129, 130.)

Glaston (188) a village on the Stamford road, two miles from Uppingham. The church is mostly of the Decorated period. In the chancel are some monuments of early date, one of which is in memory of Walter Colley, who was lord of the manor in 1407.

Greetham (531) is a long village on the high road between Cottesmore and Stretton, six miles from Oakham. The church is in the Transitional and Decorated styles with a Decorated western tower and broach spire. The "Greetham Inn," a famous old coaching house in this parish, on the Great North Road, is now converted into a farmhouse. (pp. 29, 82, 98.)

Hambleton (241) stands on an eminence three miles south-east of Oakham; the views to be obtained from the village

Exton Church

are of a most charming and varied character. The church is an ancient fabric, much restored and enlarged from time to time. A curious and unusually moulded Norman doorway will be found at the south entrance.

The Old Hall, built in the early part of the seventeenth century, lies on the slope of the hill overlooking Hambleton Wood and Gibbet Gorse. It is a fine specimen of Renaissance architecture with Jacobean oak staircase and panelling in the interior. (pp. 19, 27, 30, 80, 94, 95, 110.)

West Doorway, Ketton Church

Ketton (992) is situated on the north side of the river Chater, on the Uppingham road, about four miles from Stamford. Connected with this village are the hamlets of Geeston and Aldgate on the south side of the river, as well as that of Kelthorpe. There are extensive stone quarries here, noted for their excellent freestone. The Hall is a handsome modern mansion in the Tudor style. The church is of Norman origin with a tower crowned by a beautiful broach spire rising to a height of nearly 180 feet. The western door is one of the finest specimens of the transitional style from Norman to Early English architecture in the kingdom. The church was in a very dilapidated state in the early part of the thirteenth century and Hugh de Welles, Bishop of Lincoln, granted a release of twenty days' penance to all who contributed anything towards its restoration. (pp. 19,

Langham Church
(*strewn with hay*)

30, 31, 39, 62, 66, 67, 69, 83, 86, 89, 95, 97, 98, 109, 112, 121, 128, 129.)

Langham (625) is a large scattered village two miles north-west of Oakham, with which it is ecclesiastically connected. On the western side is a bold eminence called Ranksborough Hill, which is one of the finest fox covers in the county. The Old Hall is a Gothic building, erected in 1665. The church is a fine specimen of Decorated and Early Perpendicular architecture said to have been built by Simon de Langham, Archbishop of Canterbury. It once contained much armorial glass, with the shields of the Kings of the East Angles, of the Beauchamps, Earls of Warwick, and of the ancient families of Hastings and Clare. The curious custom of strewing the church with newly mown hay on the Sunday after the feast of St Peter is kept up here. There is a small piece of land in the parish which was bequeathed for the purpose of supplying the grass. The custom is probably the outcome of the use of straw or rushes when places of worship had nothing but mud floors. (pp. 27, 61, 63, 83, 89, 90, 95, 98, 100, 132.)

Luffenham (North) (431) stands on the northern side of the valley through which runs the river Chater, seven miles from Stamford. The church is in the Early English and Decorated styles with a western tower and broach spire. In the chancel is a brass to the memory of Archdeacon Johnson, founder, in 1584, of Oakham and Uppingham Schools, who was rector here. There is some very fine old armorial glass, circa 1350, in the window on the north side of the chancel. (pp. 17, 19, 28, 32, 77, 86, 89, 95, 100, 132.)

Luffenham (South) (329) stands on the banks of the river Chater about a mile from North Luffenham. The church is an ancient building. The arcade on the north side of the nave is Norman, that on the south side is Early English; the lofty chancel arch is also of this period. There is an embattled tower crowned with a crocketed spire. The ancient custom of beating the bounds is kept up here on Rogation Monday. (pp. 17, 19, 62, 95, 121.)

Lyddington (366) is a long village, lying in a valley, nearly two miles south-south-east of Uppingham. The manor belonged to the Bishops of Lincoln, and Bishop Russell, who was translated from Rochester to the See of Lincoln in 1480, built a palace here between that year and 1496. This was converted into a Hospital for poor people and endowed by Thomas Cecil Lord Burleigh, son of the great Lord Treasurer Burleigh, in 1602. The banqueting hall, with its splendidly carved ceiling, open fire-place, and oriel window are worthy of special notice. The richly moulded tracery of the ceiling is a very good specimen of late Perpendicular wood-carving, and over the fire-place still remain the arms of Bishop Russell. The garden is walled and at one of the corners is a curious octagonal tower which projects over the pathway of the village street. The church is an ancient structure containing nave and aisles of late Perpendicular work. The tracery of the windows on the south side of the chancel are very good examples of the Decorated style. The old rood screen remains, together with the stairs in their original condition. The altar rails are curious, enclosing the communion table on all sides. They bear date 1653. An order in Council was made, in 1633, by Archbishop Laud making it compulsory to fence off the communion table by rails from the body of the church. The then Bishop of Lincoln was Lord Keeper Williams, in many ways an opponent of Laud. He characteristically interpreted the order to mean, and gave orders to the same effect in his diocese, that " the Holy Tables should be placed in the middle of the chancels and railed in." This arrangement, the only one found in Rutland, is historically interesting as a memento of Bishop Williams and of a practice more or less general at one time throughout his diocese. (pp. 38, 50, 89, 90, 98, 100, 110, 129.)

Lyndon (197), a small village about four-and-a-half miles south-east of Oakham. The Hall is a fine old English mansion, built in 1675, and long the seat of the Barker family. The church is a small building in the Decorated style. In the church-yard is a tablet to the memory of the Rev. William Whiston, celebrated as the translator of the works of Josephus, who died in 1752. (pp. 29, 55, 133.)

Manton (291) is a small village, on a bold eminence south of the river Gwash, about three-and-a-half miles from Oakham on the Uppingham road. The Midland railway passes through a tunnel about half a mile long under the hill. The church, which is of Norman and Early English architecture, has a beautiful transitional Norman bell-turret containing two bells. (pp. 8, 18, 62, 80, 95, 100, 121, 135.)

Market Overton (403) is pleasantly situated on a height on the borders of Leicestershire about six miles north-east of Oakham. The church stands within the boundary of a Roman camp. The tower arch is the only specimen of Anglo-Saxon work in the county. Many Roman remains have been unearthed in the parish, consisting of coins, steelyards, fibulae, bronze and bone implements in great variety, and many specimens of Samian ware. On the exterior of the church tower is a sundial, said to have been given by Sir Isaac Newton, whose grandmother lived some years in the village, and whose mother was born here. (pp. 8, 14, 28, 49, 50, 65, 66, 80, 82, 83, 84, 86, 87, 100.)

Morcott (392) lies on an acclivity above one of the tributary streams of the river Chater, four miles from Oakham on the main road between that town and Stamford. The church dates from the twelfth century. The nave is the most perfect example of Norman work to be found in the county and is in excellent preservation. There is an ancient monument inscribed in Norman French to the founder, William de Overton. (pp. 17, 28, 61, 95, 121.)

Normanton (64) is on the south side of the river Gwash, about seven miles west of Stamford. The Hall (the seat of the Earl of Ancaster) has been described in the chapter on Architecture. It stands in a beautiful park of about 500 acres, which is stocked with deer. On the outskirts are some large workshops with wood-working machinery where upwards of 100 workmen are employed. The church is a small building in the Romanesque

Market-place, Oakham

style, having a portico of the Ionic order, with richly carved Corinthian columns above, forming a bell-turret. There are a few mural monuments to the Heathcote family and one to the Baroness Willoughby de Eresby. (pp. 9, 40, 43, 112, 138.)

Oakham (3667), the county town and the centre of county business, is pleasantly situated among the green pastures of the fertile Vale of Catmose. The houses that line its long main street and market square are old fashioned. Some even are mere thatched cottages. There are a few examples of the stone-mullioned windows and gabled oriel characteristic of the district. The market-place, grass-grown in places, with its octagonal Butter Cross, a wooden structure covered with a lofty pyramidal roof raised on massive posts and sheltering the town stocks, offers, with the tower and spire of the church in the background, a charming subject for the sketch-book. But the chief objects of interest in Oakham, well worth making a journey to see, are the Castle Hall, and Flores House, and the beautiful church, All Saints. The Castle Hall is elsewhere described, so we may pass on to Flores House. The Flores or Flowers were the leading family at Oakham for several centuries, their name constantly appearing in the records of the town and its charitable foundations, and to one of them the beautiful fourteenth century spire of the Church is attributed. The mansion known as Flores House, the picturesque front of which is one of the chief ornaments of the main street, preserves, amid much alteration, several Early English features. The house consisted of a hall in the middle flanked by gabled wings projecting backwards. It stands at the western entrance of the town, which was never walled, the street here being narrowed to the width of one vehicle. Here, as elsewhere, it may be observed how in medieval towns the wider thoroughfares of the inner part of the town were only approachable by passages of very contracted dimensions, and usually with sharp turns admitting of easy defence against

assailants.   The principal entrance of the house is a very good
example of a shafted doorway of the thirteenth century, one of

All Saints, Oakham

the dripstone terminations being a crowned head with the short
beard and hair of Henry III.   Within, on the end wall of the

hall, is a curious early water-drain, resembling a piscina, for washing the hands before and after dinner, an arrangement of which later examples are found at South Wingfield in Derbyshire, Dacre Castle, and the Deanery, Wells.

The Church of All Saints, which stands to the west of the Castle, is a singularly beautiful building both in outline and details. The restoration in 1858, while productive of much

Oakham School

good in clearing away pews and galleries and allowing the fine proportions of the interior to be seen, appears to have done no more harm than is inseparable from the work of the restorer. The western tower, with its spire of Late Decorated date, is a very stately composition. The relative proportions of the successive stages are excellent; while the junction of the spire is admirably masked by the unusually large corner turrets which take the pinnacles. The general effect of the exterior is of a building of Perpendicular date, but, as so often happens, on entering we find that later walls and windows mask an earlier fabric. The nave arcade is Early Decorated work to which very wide chantry aisles have been added of Late Perpendicular reaching quite to the east end. There are quasi-transepts formed by doubling the two eastern bays of the aisles, under a flattish gable. The general effect of the interior is one of singular stateliness and spaciousness. There is a fine tub-shaped Norman font, elevated on a base which was probably that of a churchyard cross.

At the north-east corner of the churchyard is seen a long high-pitched gabled building of about 1584, the original building of the celebrated Grammar School founded by Archdeacon Johnson, who also established the sister foundation at Uppingham. The room is now used as a museum and accommodates the first-form boys. Beautifully executed frescoes have been recently painted on the walls illustrating the Story of Gareth. A figure of the founder is on the east wall and on either side a list of the Exhibitioners since 1710. The School House, a modern building of stone, is in the market-place. The buildings are suited to modern public school requirements and have accommodation for about 120 boys. The juniors are provided for by a recently erected house overlooking the playing fields.

The trade of Oakham consists chiefly in cattle, for which weekly markets and monthly fairs are held. There is a small shoe-factory employing about 100 workpeople, and the Midland Railway Company has one of their huge forage stores here.

The recently issued (1911) census return shows that the population has increased during the decade by 374, which accounts for 58 per cent. of the total increase in the county. (pp. 5, 9, 15, 27, 42, 47, 50, 53, 55, 61, 62, 63, 69, 70–73, 75, 80, 89, 90, 98, 100, 102–105, 107, 108, 114, 117, 119, 121, 122, 128, 138.)

Pickworth (154) is a small village five-and-a-half miles north-north-west of Stamford. It is said to have been once of considerable size, but the ancient church went to decay some centuries ago and there now remains only one of the arches of the porch, supported by combined columns having richly foliated capitals and moulded bases. The church and village which surrounded it are supposed to have been destroyed by the rebels after the Battle of Losecoat Field. The present church, which was erected in 1824, contains no features of interest. (pp. 31, 32, 137.)

Pilton (36) lies on the south side of the valley of the river Chater four-and-a-half miles north-east of Uppingham. There is nothing interesting about the church except that it has one of the six Rutland examples of a double bell-cot. The chancel was rebuilt in 1852, and the other parts of the building restored in 1878. (pp. 28, 90, 100.)

Preston (243) stands on an eminence two miles north of Uppingham. It contains several specimens of seventeenth century buildings with stone mullioned windows and high-pitched gables giving a picturesque character to the village. The church, which appears to have been mostly rebuilt in the fourteenth century, affords examples of almost every style, from Norman down to Renaissance. The nave and aisles are of Norman character, and exhibit the chevron and zigzag ornament, but the chancel and tower arches are pointed, the former being about the date of 1230. There is in the church an interesting series of parish records, chiefly relating to taxes and rates levied during the Civil War and Commonwealth. Copies of many of the documents have been deposited in the British Museum. (pp. 8, 18, 29, 40, 71, 95, 113.)

Ridlington (221), a village on the south side of the river Chater three miles from Uppingham. It stands within the ancient limits of Leighfield Forest and in the midst of rich woodland scenery. The oldest parts of the church are the tower and the arcade on the south side of the chancel arch, which are thirteenth century, Early English work. Over the vestry door is preserved a curious tympanum of a Norman doorway, portraying a griffin and a lion, with an eight-spoked wheel within a circle. The symbolism is very obscure but probably is intended to convey the lesson of the never-ending conflict in this world between good and evil. (pp. 18, 29, 33, 39, 50, 62.)

Ryhall (701) including Belmisthorpe. This village stands on both sides of the river Gwash about three miles north-east of Stamford. In the seventh century, St Tibba, a kinswoman of Peada, King of Mercia, is said to have dwelt in a cell or chapel formerly attached to the west end of the north aisle of the church and was buried there. Ingulphus, abbot of Crowland, who lived at the time of the Conquest, says that in the bloody battle fought by the Danes in 870, the stout knight, Harding of Ryhall, was one of the commanders of the Stamford men. The church is in the Early English style with western tower and broach spire. In the chancel are two monuments of the Bodenham family dated 1613 and 1671. (pp. 20, 32, 71, 89, 95, 100, 109, 122.)

Seaton (198), a village standing on the north side of the river Welland, two-and-a-half miles from Uppingham. The church is of the transitional Norman and Early English styles. The inner doorway is Norman, twelfth century work, with moulded arch and richly carved shafts. The fine chancel arch is of the same date. (pp. 18, 29, 39, 86, 89, 95, 98, 119, 121.)

Stoke Dry (61) is a small village two miles west of Uppingham, situated on the rising ground on the east side of the valley of the Eye river, which separates it from Leicestershire. It commands extensive views over the valleys of the Eye and Welland, beyond which is seen a portion of the ancient forest of Rockingham, with its Norman castle in the foreground. The manor once belonged to the Digby family, of whom Sir Everard was one of the Gunpowder Plot conspirators. The church is small but very ancient. The chancel arch is Early English, supported on slender Norman pillars richly carved with human figures, animals, foliage, etc. There are several very interesting monuments of the Digby family here, including the figure of a knight in armour representing Everard, who died in 1440, and a table monument with effigies of Kenelme Digby and Anne his wife and their eleven children. (pp. 8, 18, 29, 40, 78, 95.)

Stretton (170) lies on the east side of the Great North Road about eight miles from Stamford. An old Rutland proverb designates it " Stretton-in-the-Street, where Shrews meet." Stocken Hall, on the parish boundary about two miles away, is a fine old mansion, built in the reign of Charles I. The church is an interesting fabric of an Early English character with a double bell-cot. Edward Bradley (" Cuthbert Bede "), author of the *Adventures of Mr Verdant Green*, etc., was rector here from 1871 to 1883. (pp. 32, 82, 95, 100, 116, 118, 130, 136.)

Teigh (110) is a small village five miles north of Oakham. The church was rebuilt in 1782, except the tower. The interior is a curious mixture of Grecian and Gothic. The pulpit and reading and clerk's desks are in the western arch, and their arrangement is quite unique. There are neither aisles nor chancel and the only entrance is under the pulpit, from the interior of the tower. (pp. 27, 28, 40, 100, 137.)

Thistleton (114), a village near the junction of Rutland with Leicestershire and Lincolnshire, eight miles north-east of Oakham. The church was rebuilt in 1780 in the Gothic style. A Roman camp is supposed to have existed here. A large number of Roman coins and other antiquities have been unearthed in the neighbourhood, and a Roman well. (pp. 12, 30, 82, 83, 84, 86, 119.)

Tickencote (110), a village, with a large water-mill, on the north side of the river Gwash, three miles from Stamford. The church is of very early origin and was probably built by Robert

Grimbald soon after the Conquest. It was rebuilt in 1792, special care being taken to preserve the chancel, which is one of the finest specimens of Norman work extant. The chancel arch consists of five recessed divisions, covered with elaborately carved figures characteristic of the later Norman period. In the chancel is part of a wooden effigy in armour supposed to represent Sir Rowland Daneys who died in 1362. (pp. 20, 90, 94.)

Tinwell (218) is picturesquely situated on the north side of the river Welland on the Uppingham road, about a mile-and-a-half from Stamford. The church is a small structure partly of Norman date. The tower has a high pitched saddle-back slated roof, the only example in the county. There is a monument in the church to Elizabeth, sister of the Lord Treasurer Burleigh. (pp. 18, 19.)

Tixover (49), a small village on the north bank of the Welland six miles south-west of Stamford. The church, which is some distance from the village, is a fine specimen of Early Norman, Transitional, and Early English work. The Early Norman tower is nearly in its original condition except for the embattled parapet, which was a later addition. There is a fine marble monument, with kneeling figures, in the north aisle, to Robert Dale and his wife. He was lord of the manor and died in 1623. (pp. 18, 39, 90, 95, 129.)

Tolethorpe, a hamlet in the parish of Little Casterton. There is here a curious old mansion, in the Elizabethan style, but partly modernised. It stands on an eminence overlooking the river Gwash. There is said to be a chalybeate spring near the house, possessing properties similar to those of the waters of Tunbridge Wells. (pp. 20, 102, 110.)

Uppingham (2573), the only other town besides Oakham in the county, consists, principally, of one long street stretching along the very edge of one of the long, low, steep hills which are

P. R.   11

Uppingham Church

characteristic of this part of Rutland. The town does not appear of such importance as to be mentioned in Domesday Book. There is no particular notice of it until 1265, when Sir Peter de Montfort, one of the retinue of his kinsman, the great Simon de Montfort, Earl of Leicester, the founder of the House of Commons, both of whom with many other barons fell in the Battle of Evesham in 1266, gave it to his second son William de Montfort. In 1400 we find it in the hands of Thomas, Earl of Warwick, a well-known character in English History. Richard Neville, Earl of Warwick, known as the "King Maker" became possessed of the manor of Uppingham in 1449. In the early part of the sixteenth century the manor was alienated to the Crown and King Edward VI granted it to his sister Elizabeth, who, when she became Queen, gave it to the Earl of Exeter. A wealthy citizen and mercer of London, named Edward Fawkener, purchased the manor, and it remained in his family until it became the property of the Noel family, the present holder being the Earl of Gainsborough.

The town has some good stone-fronted houses, both in the earlier mullioned and gabled style and in that of the first Georges. The neighbourhood abounds in beautiful scenery and there are fine views from the western side of the table-land over the Leicestershire hills, and from the south and south-east over the Welland valley. The most interesting object of antiquity in the parish is a mound called Castle Hill (see p. 15), situated near the Leicester road, about a mile from the town, bearing the remains of military works, and commanding a splendid view of Deepdale and Beaumont Chase.

Beyond its having been the first incumbency of Jeremy Taylor, and having sheltered Charles I on his flight from Oxford to Southwell, on the night of Saturday, May 2nd, 1646, the annals of Uppingham are bare of historical incidents. The church, which stands conspicuously on the brow of a steep descent, at one end of the market-place, though much inferior to many churches in the county, is far from deserving Leland's contemptuous description as a "very meane churche." Originally of Norman foundation it was entirely rebuilt, in the time of Edward I, in the Early Decorated style, and has a well-proportioned western tower and spire, a conspicuous object from all the country round. The interior was unfortunately restored in 1860. The chief object of interest is the old Jacobean pulpit, from which Jeremy Taylor preached when he was rector of Uppingham. This, though robbed of its sounding-board, and

Uppingham School

otherwise meddled with, is as it was in Taylor's days, and is shown to the visitor as "Jeremy Taylor's pulpit" (see p. 135).

The most marked feature in Uppingham is the group of school buildings, and the chapel, on the south side of the High Street. The chapel is in the Early Decorated style. The east window, of five lights, is filled with stained glass, depicting the scenes from the death and passion of Christ, and is the gift of old boys. The west end contains an excellent example of a rose or wheel window. There are several monuments, notably

Edward Thring

the life-sized figure, in marble, by Sir Thomas Brock, R.A., of the late Rev. Edward Thring, the maker of modern Uppingham. Uppingham School, as well as the sister foundation at Oakham, was originally established and endowed, in 1584, by the Rev. Robert Johnson, Archdeacon of Leicester and Rector of North Luffenham, as a free Grammar School for poor boys. The ancient foundation was reconstituted by the Charity Commission and under the able management of the late Mr Thring, Uppingham increased from 25 to nearly 400 boys, and takes rank among the leading public schools of the country.

The original school room, of the same humble type as that of Oakham, still stands at the south-east corner of the churchyard, but has been converted into a drawing-school.

A small market for cattle is held weekly; there are no manufactures, the town depending principally on the school for its trade. Uppingham is a terminus of the L. & N. W. R. on their branch line from Seaton. (pp. 10, 16, 29, 61, 89, 95, 98, 100, 119, 121.)

Wardley (43) is a small village which lies south of the Leicester road two-and-a-half miles from Uppingham, a picturesque place of hill, dale, and wood. The church is an ancient structure in the Early English style and contains several monuments of the Fludyer family. (pp. 8, 29, 40.)

Whissendine (673), a large village in a hilly district five miles north-west of Oakham. The church is in the Early English, Decorated, and Perpendicular styles, with a very fine Decorated western tower. There is here a handsomely carved oak screen, erected between the south chantry and aisle, with a richly groined canopy. This screen originally belonged to the chapel of St John's College, Cambridge, to which it was presented by Margaret, Countess of Richmond, mother of Henry VII. (pp. 8, 12, 20, 26, 61, 89, 95, 98, 100, 119, 121.)

Whitwell (78) is on the Oakham and Stamford road four-and-a-half miles from the former town.   The church is a small but ancient structure in the Early English style with double bell-turret.   (pp. 28, 50, 95, 100.)

Wing (297), a village on the south side of the valley of the river Chater, four miles from Uppingham.   On the east side of the village is an ancient turf maze, which has been carefully preserved, and is recut periodically.   The church is an ancient building, supposed to have been built in 1335, on the site of a Norman structure.   The doorway is of the late Norman style, the tower Perpendicular.   (p. 80.)

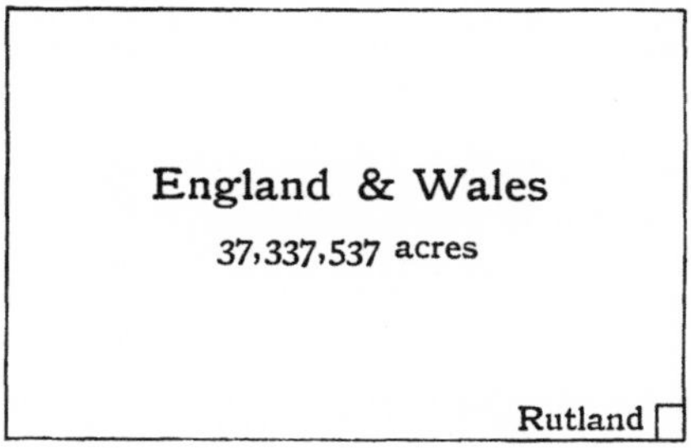

Fig. 1. The Area of Rutland (97,273 acres) compared
with that of England and Wales

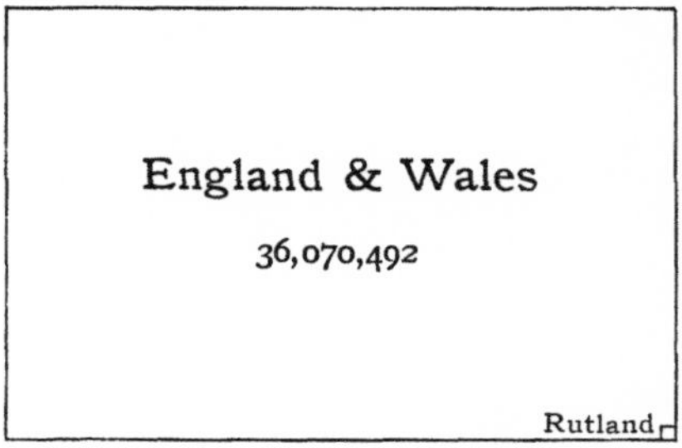

Fig. 2. The Population of Rutland (20,346) compared
with that of England and Wales in 1911

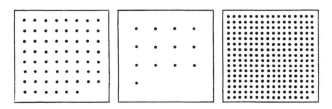

England and Wales 618   Rutland 134   Lancashire 2554

Fig. 3. Diagram showing comparative Density of Population
to the square Mile

(*Each dot represents ten persons*)

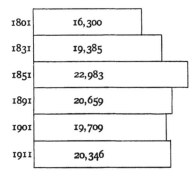

Fig. 4. Diagram showing Increase and Decrease of
Population in Rutland since 1801

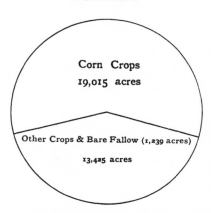

Fig. 5. Proportionate Area under Corn Crops compared with that of other cultivated land in Rutland in 1911

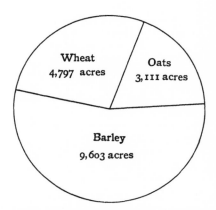

Fig. 6. Proportionate Area of Chief Cereals in Rutland in 1911

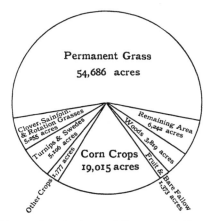

Permanent Grass
54,686 acres

Clover, Sainfoin, & Rotation Grasses 5,255 acres

Turnips & Swedes 5,106 acres

Other Crops 1,777 acres

Corn Crops
19,015 acres

Remaining Area 6,242 acres

Woods 3,849 acres

Fruit & Bare Fallow 1,373 acres

Fig. 7. Proportionate Area of Permanent Grass to other
areas in Rutland in 1911

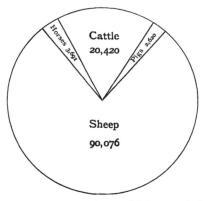

Horses 3,691

Cattle
20,420

Pigs 2,620

Sheep
90,076

Fig. 8. Proportionate numbers of Live-stock in Rutland
in 1911

www.ingramcontent.com/pod-product-compliance
Ingram Content Group UK Ltd.
Pitfield, Milton Keynes, MK11 3LW, UK
UKHW042144280225
455719UK00001B/81